More praise for

THE ELECTRIFYING FALL OF RAINBOW CITY

"Margaret Creighton does for Buffalo in 1901 what Erik Larson did for 1893 Chicago in *The Devil in the White City*. Creighton's book is a propulsive, edge-of-your-seat ride. . . . Creighton creates a vivid panoply of daredevils, hucksters, suffragists, and civil rights champions, as she conjures up the sights and sounds, the very aromas and tastes, of America at the turn of the last century."   —Lauren Belfer,
author of *City of Light* and *And After the Fire*

"While perhaps not quite as well known as the 1893 World's Fair in Chicago, the 1901 Pan-American Exposition in Buffalo, New York, was equally full of drama and intrigue. . . . *The Electrifying Fall of Rainbow City* is the compelling story of an event that sparked technological advances and spurred new perspectives on social equality and race."
—Becky Diamond, *BookPage*

"Electricity defined the 1901 World's Fair, and Creighton's readers become omnipresent fair-goers. Attending this illuminated extravaganza at the start of the new century, we absorb the pomp and pageantry, travel backstage to enter the lives of impresarios and performers, witness the assassination of President McKinley, and meet the assassin. This is a story of American democracy and American imperialism, of modern wonders intertwined with disillusionment and exploitation—all unfolding in utterly electrifying prose."   —Martha Hodes,
author of *Mourning Lincoln*

"[*The Electrifying Fall of Rainbow City*] is primed to join the ranks of the best history books to be written about Buffalo."
—Mark Sommer, *Buffalo News*

"An extraordinary portrait of the event . . . great storytelling and painterly in its color and detail." —Mark Goldman,
author of *High Hopes* and *City on the Edge*

"In her electrifying account of this electrifying fair, Margaret Creighton reveals how the 1901 Pan-American Exposition heralded the dawn of the 'American Century.' This is a highly readable and insightful history of a fair brightened by the latest in technological marvels and darkened by the assassination of an American president."
—Robert W. Rydell, author of *All the World's a Fair*

"Captivating . . . the chapters read with the high drama and attraction of fiction." —James A. Cox, *Midwest Book Review*

"I was born in Buffalo, and the Pan-American Exposition was very much a part of my family's lore. Margaret Creighton's work is a great pleasure to read because of its wealth of incident and detail and the author's abiding fondness for the spirit and optimism that pervaded so much of the main event. Furthermore, she ends her work with an equally compelling look at Buffalo today, as the gallant old city attempts to pull itself up and reassert its ancient energy."
—A. R. Gurney, playwright and novelist

"Fascinating and horrifying." —Rob Caldwell, WCSH TV

"Gripping. . . . Creighton is a gifted writer."
—Nicholas Evan Sarantakes

OTHER BOOKS BY MARGARET CREIGHTON

The Colors of Courage:
Gettysburg's Forgotten History

Iron Men, Wooden Women:
Gender and Seafaring in the Atlantic World, 1700–1920
(edited with Lisa Norling)

Rites and Passages:
The Experience of American Whaling, 1830–1870

Dogwatch and Liberty Days:
Seafaring Life in the Nineteenth Century

THE
ELECTRIFYING
FALL OF
RAINBOW CITY

SPECTACLE AND ASSASSINATION AT
THE 1901 WORLD'S FAIR

Margaret Creighton

W. W. NORTON & COMPANY

Independent Publishers Since 1923

New York | London

For information about permission to reproduce selections from this book, write to
Permissions, W. W. Norton & Company, Inc., 500 Fifth Avenue, New York, NY 10110

For information about special discounts for bulk purchases, please contact
W. W. Norton Special Sales at specialsales@wwnorton.com or 800-233-4830

Manufacturing by Quad Graphics Fairfield
Book design by Ellen Cipriano
Production manager: Louise Mattarelliano

Library of Congress Cataloging-in-Publication Data

Names: Creighton, Margaret S., 1949– author.
Title: The electrifying fall of Rainbow City : spectacle and assassination at the
1901 World's Fair / Margaret Creighton.
Description: First edition. | New York : W. W. Norton & Company, [2016] |
Includes bibliographical references and index.
Identifiers: LCCN 2016018256 | ISBN 9780393247503 (hardcover)
Subjects: LCSH: Pan-American Exposition (1901 : Buffalo, N.Y.) | Pan-American
Exposition (1901 : Buffalo, N.Y.)—Biography. | Exhibitions—Social aspects—New York
(State)—Buffalo—History—20th century. | Spectacular, The—Social aspects—New York
(State)—Buffalo—History—20th century. | McKinley, William, 1843–1901—Assassination.
| Scandals—New York (State)—Buffalo—History—20th century. | Buffalo (N.Y.)—Social
life and customs—20th century. | Buffalo (N.Y.)—Social conditions—20th century.
Classification: LCC T485.B2 C74 2016 | DDC 607/.3474797—dc23
LC record available at https://lccn.loc.gov/2016018256

ISBN 978-0-393-35479-9 pbk.

W. W. Norton & Company, Inc., 500 Fifth Avenue, New York, N.Y. 10110
www.wwnorton.com

W. W. Norton & Company Ltd., 15 Carlisle Street, London W1D 3BS

1 2 3 4 5 6 7 8 9 0

For Jean Scott and Dotty
Best friends and Buffalonians

Contents

Rainbow City at night: the view from the Esplanade.

THE
ELECTRIFYING
FALL OF
RAINBOW CITY

Prologue

1901

I t was closing in on dusk that November day when people in the stadium, who had been shivering in the cold, were finally rewarded with a show. They could see movement in the shadows—gray shapes in the gray light—and soon they could make out the animals. Three small elephants moved into the chasm of the arena and, with them, a giant hulk. He was the one they waited for—Jumbo II, the largest mammal in captivity.

The spectators stilled as the big elephant came to a stop. He ran his trunk lightly around the smaller animals, toying with them. The crowd quieted further when Jumbo's owner, a big, mustached man with the moniker of The Animal King, stepped forward. In his tailored English, the King recounted Jumbo's history, from his decorated army career in India to his long Atlantic crossing. He spoke of the elephant's hard life at the Pan-American Exposition, which had just closed. Jumbo had been at the fair only a few months, the King explained, but it was clear he would never adjust to show business. He was disobedient and had a temper. Although he had not killed anyone, he might do so at any time. He must be destroyed.

The elephant's quiet presence may have unnerved some of the audience, and his playful attentions to the smaller animals may have stirred some doubt. But no one left. They wanted to see the old war veteran electrocuted.

The shock would be delivered by voltage from Niagara Falls, twenty-six miles away. If the onlookers had been to the big Buffalo exposition at all over the previous six months, they would know how apt this moment was as an end to the fair. The Pan-American Exposition had been all about electricity. The most talked-about building was the Electric Tower, which, in homage to its wondrous source of power, gushed out a miniature waterfall. Electricity ran the fair's generators and transported guests via street railways. Its alternating current turned on thousands and thousands of tiny lights at night, sending visitors into raptures.

For some, then, it was fitting that Jumbo II be shocked to death. The event would be a supreme act of Western accomplishment: It would harness a natural wonder; it would control a mysterious power; and it would bring down one of Asia's biggest beasts.

The elephant's demise would echo some of the most popular shows on the Midway, too. On the Lane of Laughter, as it was called, showmen prodded elks into high-water dives and goaded apes into theatrical performances. The Animal King himself had shipped a veritable ark to Buffalo. His press agent claimed that every beast—from snakes to lions to bears to monkeys—did his bidding. Before him, the agent asserted, "animals cower."

The victory dance that the Exposition performed over nature made the Pan-American fair unique. But it was not the only act in its global show of power. Like other fair promoters at the turn of the twentieth century, Pan-American directors celebrated their idea of civilization. Borrowing Darwin's vocabulary (and eschewing his science), they staged a spectacle of development, where, at every turn, they taught fairgoers about which sort of humans had advanced and which had not. They chose the progress of the Western Hemisphere as the theme for the fair and signaled how far some nations—namely, the United States—had come and how others—namely

Latin American republics—labored to catch up. Art directors had even applied these ideas to the tints they chose for exhibition buildings. At the southern entrance to the show, painters coated exterior walls in "barbaric" colors and brushed structures to the north with whiter tints. The pinnacle of civilization, at the northern end of the grounds, was the ivory-hued Electric Tower. The color scheme of the fair earned it the name of "Rainbow City."

Visitors were urged to hurry to Buffalo, to take a last look at a vanishing world. This might be a final chance to meet the Apache Geronimo, who had been brought all the way from his Oklahoma prison. Plains Indians had traveled east by rail from faraway reservations and were reenacting battles and losing the West again, show after show. Filipino warriors had arrived, too, and were living and performing in a miniature Native village. The Philippines was such a new possession that even as the fair opened, American soldiers in the distant Pacific were still killing and dying to assert their sovereignty.

And visitors were told to see Niagara Falls before it disappeared. Word had it that even the mighty world wonder could be threatened. Engineers had been drilling tunnels through its broad cliffs to produce the miracle of electric power. But what would be the result? The thunderous cataract might be reduced to a rock ledge and a trickle of water.

And seeing Jumbo II, the big elephant, was a must. There he stood, motionless in the dark stadium, as attendants circled around him with wires. One man in the crowd thought he looked like a grand piano, the way his legs were sprawled and tied. The Animal King claimed that the elephant knew something was about to happen: "He's crying like a baby," he said.

The King may have been right. Maybe Jumbo did sense something was up. Maybe the elephant sensed that everybody there, including the man at the power switch, was in for a big surprise.[1]

The Pan-American Exposition that featured the veteran elephant opened its gates in May 1901. Buffalo joined an illustrious set of cities that had sponsored similar events, during what would become known as the grand era of world's fairs. Begun in 1851 with London's Crystal Palace, expositions had, by the 1890s, become colossal social magnets. Those located in Western Europe and the United States flaunted military and industrial power, new technologies, and consumer goods. They became extravagant advertisements for nation states, and, when possible, showcased colonial possessions. Produced by men with big egos and fat wallets, they became sites for superlatives, offering visitors the latest, the best, and the company of the most famous.

In 1901, Buffalo hoped it had the formula down. It would borrow displays from earlier fairs, add a novel theme and design, spend big money on a midway, and count on visitors—millions of visitors. The Queen City of the Lakes, as it was known, also offered visitors Niagara Falls, just a trolley ride away. Some folks in this determined metropolis dreamed that the Pan-American Exposition of 1901 would be the biggest fair of the age, more popular even than Chicago's extravaganza in 1893. The United States had pulled itself out of a grinding recession, and more people had pocket money. There was the new century to celebrate, too. Rainbow City would bring the country into the 1900s with dazzle and pomp.

If it performed as promised, the Exposition would also allow Buffalo to finalize its grand potential. The city had always had ambition, it seemed, but needed a push to achieve greater prominence. Maybe it was playing second fiddle to New York City, the urban behemoth to the southeast, that humbled it time and again, or the fact that the halcyon days of the Erie Canal had passed. Or perhaps the challenge rested with residents who believed their big city was more of a way station than a destination. As one of them put it, Buffalo was "such a convenient stopping-off place from North to South and East to West."

In 1901, city leaders had no patience for such humility. It was time for Buffalo to think big.

When it debuted in May 1901, the Pan-American Exposition portended nothing but success. Despite wet weather, hundreds of thousands of people flocked to western New York in early summer and praised the fair's architecture and fountains, its sparkling tower, and especially its nighttime illumination. They marveled at the latest wonders of American might and ingenuity, and, for the very first time, came face-to-face with exhibitors and guests from Cuba, Mexico, and even Argentina and Chile. They found the ridiculous, raucous Midway irresistibly charming. The Exposition attracted such enthusiastic patrons that some local fans trekked to the fair over and over again. One Buffalo schoolteacher walked through the fair's front gates on more than thirty separate occasions.

There were critics, of course. Some observers balked at the fair's design. Others were underwhelmed by the color scheme. And there were people on display, in the Midway, who acted out. They protested their part in the show, sometimes publicly and sometimes quietly, in ways that only those with eyes or ears alert to such things would notice. At least two of them had come to the United States not to celebrate the nation's imperial plans but to undermine them, by force.[2]

But, looking back, these were all minor disruptions—nothing compared with what happened in the fall. It was in September, as the weather cooled, that desperation took hold. It possessed a slight, brown-haired man who had come to the fair from Ohio. Compelled by illness and personal pain, this man, who went by the name of Fred Nieman, signed into a Buffalo lodging house. Laid off from a factory job, Nieman saw himself as a casualty of the country's industrial progress. Too few people in America were wealthy, he thought, and too many were poor. He also hated the country's recent push for empire. Unlike millions of others who held these beliefs and protested them with their votes, Nieman was succumbing to a compulsion to do more. He focused his mind, his increasingly sick mind, on the American president. In September 1901, he began to stalk him.

Later on that fall, others at the fair, acting out more personally, hoped to take limelight instead of a life. One of them was a middle-aged woman named Annie Edson Taylor. As desperate as Nieman but dangerous in a different way, Taylor shipped an enormous barrel from her home in Bay City, Michigan, to Niagara Falls. The big cask, she hoped, was her ticket out of poverty. She planned to use it to carry her over the cataract and dazzle Exposition crowds. The odds were not in her favor. Nobody had ever survived such a descent.

Annie Taylor was a big woman—thick-bodied and tall. Espiridiona Cenda, also known as Alice Cenda, and most famously as Chiquita, was just the opposite. At just over two feet tall, the twenty-three-year-old Mexican was touted as the tiniest woman in the world. She was not only a Midway moneymaker; she had been crowned "mascot" of the Exposition. One day in late October, though, her English manager—the same Animal King who owned Jumbo II—said he heard cries from her quarters and found her gone. Making up a story, he told people that she had been kidnapped.

Finally, there was the elephant. Problems started, officials said, when Jumbo began acting up in October. The heavily chained animal became obstreperous, and, not long afterward, the Animal King, still reeling from Chiquita's disappearance, made the decision to try to kill him.

These events, magnitudes apart in notoriety and impact, were carried out in very different ways—from disobedience to near-suicide to escape to murder. Some acts were repugnant; others praiseworthy. Two of the actors may have been mad, two others simply pushing for freedom, one step at a time. All of them, though, big and small, offered a rebuttal to the grand Exposition. They turned the fair, with its chest-thumping displays of sovereignty over people, animals, and the earth itself, into a shocking new show. The fair would indeed bring Buffalo onto the international stage, but in an astonishing manner. And it would herald the new century—but not in the way anybody expected.

1.

Rainbow City

I
SELLING BUFFALO

Well before there was a fall season in 1901, there was, of course, a spring. In the springtime, nobody in Buffalo knew the name of the brown-haired man, or the big woman or the little one, or the name of the elephant. In their innocence, they were simply optimistic, even giddy, looking ahead to hosting one of the biggest shows on earth.

They had a hard act to follow, though, and they knew it. For all their eagerness, in fact, Exposition organizers could not escape the name of another city: *Chicago*. From the day the Pan-American fair was conceived in 1895, to the year it was supposed to start in 1899, to the day it really began in May 1901, Chicago's Columbian Exposition stood over them like a beacon and a taunt. There had been other big fairs, of course: London's Crystal Palace and Philadelphia's Centennial, both decades earlier. And in 1900, Paris's Exposition Universelle had launched the new century.

But Chicago's world's fair in 1893 had been an indisputable triumph, and Buffalo's leaders dreamed that they might match Chicago's numbers, or its style and beauty, or at least be compared to Chicago in a laudatory way.

It wasn't as though the "White City," as Chicago's exposition was known, had gone off perfectly. One of its chief architects had died before the plans had been fully realized, and the fair had opened unfinished. Then, at the very end of the exposition's run, the city's mayor had been shot and killed. To make the scene more desolate, a serial murderer had operated at the perimeter of the grounds.[1]

Chicago's fair proved to be a lot bigger than these sad events, however, and it garnered unprecedented praise. The White City, people said, inspired new industry, technology, and trade. It generated national discussions about modern architecture and art and city planning. And it had recorded more than twenty-seven million visits. Yes, some would confuse the number of visits with the number of people, and make preposterous claims, but it didn't really matter. Any way you added it up, Chicago had been breathtaking.

Omaha, Nebraska, had held the most recent fair, in 1898. Known as "The Little White City," it borrowed Chicago's colors. In fact, Omaha might have used an even more dazzling tint, for, according to *Harper's Weekly*, its neoclassical buildings—another nod to Chicago— were shockingly white. The magazine reported that men and women stood "stupefied at the entrance of the Grand Court, blinded as they would have been by a flash of lightning."

The Little White City was a small fair—drawing close to three million admissions. But it had been held during the Spanish-American War, when potential tourists were distracted, and it still had paid dividends of 92.5 percent to its stockholders.[2]

Beat that, Buffalo.

One night late in January 1899, forty Buffalo businessmen sat down at tables covered in linen and crystal, put roses in their lapels, and gathered for a do-or-die banquet. They had nursed the idea of a Pan-American Exposition for several years—at least since the Atlanta Exposition in 1895—but it was now time to give it up or go forward. Adding urgency to the occasion was the fact that Detroit also had hopes for a Cadillac Exposition, and expected to raise $500,000 within the week. If Buffalo wasn't able to top the Michiganders and go to Washington with a strong show of support, the Detroit "boomers" would have their show and Buffalo would slide back into its daily humdrum.

At that moment, a genial, philanthropic iron manufacturer named Frank Baird, an organizer of the banquet, came up with a plan. He would encourage one exuberant citizen to stand, and, in a grand gesture, pledge money to buy stock in the fair. Others, stirred by manly honor and civic duty, would leap to their feet and do the same. "The thing," Baird predicted, would "take off like wildfire."

And it did. In less than four hours, rich men stoked with enthusiasm pledged almost $500,000. A week later, the group met at the exclusive Ellicott Club for a "smoker," listened to a regimental band play a march, lit three hundred clay pipes, and cheered the fact that their fellow citizens had responded in kind. They had promises of more than a million dollars.

Most of the funds came from wealthy men, to be sure, but newsboys, firemen, artisans, laundrymen, butchers, and factory laborers invested money, too. The Polish community chipped in; the police chief put in a hundred dollars; a three-year-old girl named Esther Wedekindt sent in a bag of pennies.

Buffalonians pinched themselves. The "slumbering city," as one exultant resident put it, woke up. Another claimed the city was "newly discovered." A third insisted that "people seem to be holding their heads higher."

Mayor Conrad Diehl, an earnest, hardworking physician, was beside himself. As he arrived at the smoker he declared it was a "hot time in the new town." Diehl, who had backed the exposition idea since taking office, told the assembled men that city children had been following him on the streets all week, saying, "There goes Pan American!" After he made a toast with a stein of beer to the success of the fair, men shouted and waved their pipes.[3]

A few days later, twenty-five of these men, resplendent in fur coats and huffing against the cold, waved good-bye to the mayor and boarded an overnight train to Washington. They carried with them not only the news that almost $1.2 million had been invested in the fair, but also what they hoped was a winning idea. The Pan-American Exposition would honor the progress of the Western Hemisphere. It would appeal to empire builders by drawing on the Monroe Doctrine: No Old World countries would be permitted to install formal exhibits. Latin American republics, which would be encouraged to sponsor exhibits, could show northerners something of their cultural achievements, but above all they could demonstrate that they were ready to do business with the United States. The Buffalo delegation expected the president and Congress, not to mention investors, to salivate over the idea of new resources and new markets.

In extending invitations to Latin American countries that the United States had recently fought, annexed, or assumed some control over, Exposition promoters might naturally expect some resistance. But, as one commentator put it, the fair would send a message that, as of now, "their gigantic northern neighbor is a comrade and friend." Furthermore, another observer commented, the fair would help Latin American republics "come to know our flag better; and to know it would be to love it . . . [for] it represents all that is highest in human government and human civilization."[4]

The fur-coated Buffalonians not only hoped to sell their grand theme; they also wanted to sell their hometown. As modest as they had been two weeks earlier in each other's company, they now needed to puff out their chests. They had to remind legislators that the Queen City, boasting 370,000 people, was the eighth biggest municipality in the country. The Port of Buffalo was one of the most active shipping centers in the world, with traffic exceeding even that of the

Buffalo River with grain elevators and freighters, 1890.

Suez Canal. The city was a miracle of electrical power, having tapped Niagara Falls for trend-setting streetlights and the country's first electric streetcars. People could easily get to Buffalo, too, on the numerous trains that spiderwebbed through the region. More than half of all Americans—thirty-eight million people—could reach the city in a long day's journey or less. Throw in Canada, and forty-five million people could get there on an overnight train. No other place in the hemisphere could claim that.

The delegates also had to remind congressmen that their city had cultivated men of education and ambition, including two American presidents, Millard Fillmore and Grover Cleveland. Its businessmen had made lots of money, too, first in the Erie Canal boom and then in the railroads. Situated as they were at the very eastern end of Lake Erie and edged by the Niagara River, Erie County capitalists had lived a middleman's dream, exchanging western boatloads for Eastern trainloads.

The grain elevators that towered at the edge of the lake—which helped make Buffalo the greatest grain port in the world—spoke of their success. More recently, investors had started dreaming of steel and were planning big, belching furnaces along the lake.

The Washington delegation could also boast that Buffalo was enterprising and modern. Its tallest buildings were designed by some of the country's best-known architects, such as H. H. Richardson and Louis Sullivan. It had a green string of Olmsted parks. Its grand thoroughfare, Delaware Avenue, was arched with elms and chestnuts and anchored by imposing mansions built of every material deemed fashionable—sometimes in the same façade. And it had more asphalt streets than any city in the country. Passengers in other places suffered shattered nerves as they lurched over stone or gravel streets, but Buffalo residents enjoyed smooth, quiet rides. Throw in a turreted insane asylum—another architectural marvel—and a picturesque, Paris-inspired cemetery, and who could deny that Buffalo was a state-of-the-art city?

Lafayette Square, Buffalo, ca. 1900.

Then there was the climate. The Buffalonians made a special case for the advantages of their climate. They had, they said, enviable summertime temperatures. Industrialist Charles Goodyear went further, telling Congress that Buffalo was the "pleasantest place on Earth." While the rest of the country sweltered during the summer dog days, Lake Erie fanned cool air over the Queen City. The " lake effect," in other words, meant a delicious breeze.[5]

Congress liked what it heard. It voted to spend half a million dollars on a world-class exhibit.

II

CONFIDENCE MEN

Later on, when the hard days in the fall of 1901 caused people to look back to the beginning, they ignored all the keen, unclouded pride and found sinister signs. Men of the Indian Congress on the Exposition Midway said they should have known—they had seen a "death bird." Little Lone Wolf had spotted it—a black bird about the size of a sparrow, with pure white under its wings—sitting on the pole of his father's tepee. Distraught, he told the village. Others caught sight of it and agreed on the bad omen: A chieftain was going to die.

There were other signs, too, that things were amiss. The one big animal to die en route to the fair in April was a well-trained and costly buffalo. The same month, bicyclists were horrified to discover a body hanging from one of the Exposition entrances. Another had been thrown from the Electric Tower. When people learned that these were just dummies staged by Exposition workmen, they were relieved, but not completely. Why would anybody do such a thing?

By May 20, 1901, when the Pan-American Exposition shook off

its scaffolding and came to life, such unnerving questions had been forgotten. City officials were ready to show the world what could be done with hometown spirit, determination, and local money. Washington had paid for its big exhibit, and New York's state government in Albany had pitched in to help put up a permanent building. But Buffalo, on its own, would shoulder the rest.[6]

Exposition directors hoped for a bigger crowd than Chicago's on opening day. While President McKinley, who was in California with his ailing wife, couldn't deliver a dedication speech, Theodore Roosevelt, the popular vice president and former governor, would do the honors.

The weather worked miracles. As Dedication Day unfolded, low-lying mists lifted Rainbow City's pinnacles and spires into the air and delivered to the scene a sense of magic. Buffalo residents, who had put out flags and bunting and scoured their asphalt "as clean as a kitchen floor," cheered a parade of soldiers, officials, and performers as they moved from downtown to the fairgrounds, three miles to the north. Traveling over Main Street and then up Delaware Avenue, Mexican troops, with Mexican musicians and Mexican cavalry, marched proudly alongside United States soldiers. Carriages brought along a beaming Mayor Diehl, Exposition President John Milburn, and Vice President Roosevelt, in a tall silk hat and black overcoat. Roosevelt became so excited at the cheering crowd that he stood up in his carriage for several city blocks. On Main Street, recognizing a Rough Rider from his Cuba days, he leaned out to grasp his hand. Far in the future, this habit of standing up to greet people would make Roosevelt a perfect target for a gunman, but now he met only cheering throngs.

Almost an hour behind the military parade, Midway concessionaires wheeled up the same route with two miles worth of floats. When all the marchers had entered the grounds, and the crowds had hastened to join them, the Exposition opened with an air show. Carrier pigeons flew up and fanned out with news of the day, and gas balloons, kites, daylight fireworks, and an "eagle" flying a fifty-foot-long banner dotted the sky. A thousand more balloons rose and dropped souvenirs, and tiny shells burst open and set loose little flags.

Vice President Roosevelt joined other dignitaries to walk across the Triumphal Bridge to the Esplanade, and at midday, to loud applause, he delivered a spirited drum roll under the rotunda of the Temple of Music. Almost bouncing with energy, he spoke on familiar themes: the need for hardy manliness; the iniquities of corporate wealth; a desire for Europe to relinquish its interest in the Western Hemisphere. He also uttered comforting statements to the delegates from Latin America. "None of us," he said, "will rise at the expense of our neighbors." His most laudatory words, not surprisingly, went to his own country. It was, he said, "the mightiest republic upon which the sun has ever shone."

Roosevelt and other officials also took time to tour the fair, including the Midway. The vice president visited the Diving Elks show and pronounced the elks' steep plunge into a tub of water a "splendid act." He also stopped at Frank Bostock's animal arena, where Wallace, Jr., a lion, greeted him with a growl. Roosevelt laughed. Captain Jack Bonavita, Bostock's lion trainer, then put another stubborn lion through a routine, and hit him with the end of a whip. When the lion snarled and swatted the trainer, he was whipped again. "There's the nerve I like," Roosevelt remarked.[7]

Telegrams flew to Buffalo that day from all directions: from Ecuador and Haiti and Peru; from Canada and Cuba. William McKinley tapped congratulations from San Francisco and sent along a heartfelt message: "May there be no cloud upon this grand festival of peace and commerce," he said.

Newspapers joined in. "Today marks the birth of a new Buffalo," one announced. Within days of opening, even without Chicago's numbers, others proclaimed the fair a success. Within weeks, they became even more sanguine, announcing the "Glad News" that the Pan-American Exposition "will pay and PAY WELL!" This, they said, was based on conservative estimates. There was only one caveat. The projections were "barring accident."

Journalists across the country stirred the optimism. A reporter from the *New York World* declared the Pan-American "ought to be the most successful international exposition ever held in America." Chicago had taken place, he explained, during an economic slump, when the nation was disheartened. Buffalo would get the benefit of prosperous times, and would be able to show off eight years' worth of amazing inventions. And with Niagara Falls nearby, too? Those records were about to take a big, big tumble.

It was gratifying, too, to read how people from one end of the country to the other were planning visits to Buffalo. In Butte, Montana, several excursion groups had purchased Exposition tickets. In Pueblo, Colorado, five trains of tourists were heading east. And in a remote settlement named Old Town, Maine, nearly half the population was packing its bags. It was hard for "real" Pan-American visitors to make it, of course, but Mexican President Porfirio Diaz, it was said, had taken a "personal interest" in the Exposition, and many of his countrymen had secured tickets on boats and trains. Wealthy Cubans, too, were said to be northbound.

The only New World friends who seemed somewhat lukewarm about the show actually lived close by—within shouting distance, in fact. Canada did send plenty of visitors to Buffalo throughout the season, but its government tacitly acknowledged that the "Pan" in the Pan-American Exposition was really directed to the south, and it spent less than Chile and Cuba on its national building. In early May, an Exposition delegation traveled to Ottawa to extend a personal invitation to Canada's governor general to attend either Dedication Day or President's Day, when McKinley would visit, then scheduled for mid-

June. The governor general, Lord Minto, declared that "he would not be able to go on the twentieth, and he was also very much afraid that he could not get away on President's Day." He was unable to say when a special day for Canadians should be scheduled.

But never mind. Rainbow City drew enough praise from other quarters. Distinguished visitors delivered glowing reviews. Nikola Tesla, the famous electrical engineer, who visited the fair even before it was finished, pronounced it "grand." He would be kind, of course—his alternating current supplied the power that lighted the Exposition. But Thomas Edison praised it, too. He stopped at Buffalo in mid-July and was whisked away for a quick view of his incandescent lights at the Illumination. Taking in the nocturnal ritual, when tiny lamps grew in power until they outlined both buildings and grounds, he grew expansive. "This is the consummation of my grandest dream," he exclaimed. "It exceeds in beauty and brilliancy anything heretofore created by man." Other reports said he was more succinct. He called it "out of sight."[8]

III
AN UNINVITED MAN

Like other nearby cities, Cleveland, Ohio, put on extra trains to Buffalo that summer, and its railroads advertised specials to the Exposition. The local press cautioned, though, that it cost money to see the fair, and even more money to eat and stay near the show.

How strange it must have seemed to his family, then, that brown-haired Fred Nieman—who did not have much money, who disliked cities, and who hated so much of what the fair celebrated—would want to spend time near the Exposition. Twenty-eight-year-old Nieman, a fair-skinned man with sky-colored eyes, had spent the spring and early summer at home on the family farm. He kept a quiet routine: He did undemanding jobs, read, and slept. His father, a Prussian immigrant, was out of work, and his stepmother, brothers, and sisters labored hard to get by. Yet Nieman's family had some time ago given up hope of seeing him work hard. He was sick, he said.

Nieman did have strength enough for some pursuits. He went eagerly to political meetings and he was able to walk, half a mile, to retrieve newspapers. He devoured newspapers of all kinds—Polish, English, labor, socialist, and what he called "capitalist" papers. He fell asleep reading the papers, and then he reread them. He paid close attention to articles about working people and strikes. And he liked reading about anarchists.

Sometime in 1900, his family recalled, he read an article that provoked him or comforted him in a strange way. He clipped it out of the paper and it became so "precious" that he took it upstairs with him when he went to bed. It described the July 1900 assassination of King Umberto I in Italy, by an Italian American worker, an anarchist.

In the midsummer of 1901, Nieman's pattern of near-somnolence shifted. Displaying more energy, even a fevered intensity, he suddenly left home, boarded a train, and headed west. Then he reversed direction. By the third week of July, he had taken lodgings with a family in the town of West Seneca, a five-cent trolley ride from Buffalo. Writing to his older brother, Waldeck, he asked for ten dollars to be sent to a Fred Snider. That was a new name for him, just as Nieman had once been new. He had been born Leon Czolgosz.

Ghost-like, Nieman disappeared from his boardinghouse in the morning and returned after dark. He told his housemates that he went to meetings, but he did more than that. He went to the Pan-American fair, where he did some odd work and wandered about, looking at displays and people, drawing conclusions and calculating. He was such a calm, pleasant-faced fellow, so carefully combed and dressed, that his anger must have been easy to miss.[9]

IV
THE ENTHUSIAST

The men and women who had designed the Exposition, who had labored over drafting tables and fiddled with sketches and models, likely never pictured a man like Fred Nieman at Rainbow City. They had built the fair for people who were confident and who lived comfortably. Their ideal visitor would be someone smitten with the United States, who would take pride in the way that white Americans, with their sophisticated customs and arts, their advanced machines, and their ships and guns, were on the ascendancy. The perfect guest would be curious about foreigners, too, like the "up and coming" Latin Americans, and the "little brown men" of the Philippines, the new American possession. And the guest would lap up the curiosities and acts of the Midway, too, and buy ticket after ticket after ticket.

It would be somebody like Mabel Barnes. Of all the visitors who made their way to the Exposition over its run from May to November, twenty-three-year-old Barnes, a Buffalo second-grade teacher, was likely the most eager and most loyal. She visited the Exposition before it opened, after it closed, and nearly every week in between— thirty-three times in all.

The daughter of a Buffalo confectioner, Mabel had taken her first teaching job in the city just after graduating from high school. Now, in the summer of 1901, she dedicated her school vacation to the "Pan." Riding her bicycle or using the streetcar, she arrived alone or brought along cousins and friends. Most often, though, she went to Rainbow City with an older friend named Abby Hale. Fortified by lemonade or ginger ale, the two women circulated through grand halls and state and country exhibits, sat through concerts and lectures, cheered at pageants and parades, and glided in launches through the Grand Canal and the Mirror Lakes. Mabel took notes nearly everywhere she went and picked up hundreds of cards, free samples, booklets, and guides. She carried away souvenirs, too—a piece of petrified wood, a coconut, and, for her lucky students, a vial of bees in alcohol. She gathered so many mementos that the scrapbooks she made of her visits ballooned into a fourteen-year project.[10]

Mabel Barnes had probably never traveled to Chicago's exposition. She would have been too young, and maybe without the means to go west in 1893. But if she had read about the White City at all, she would have known it had been a neoclassical masterpiece, an elegant tribute to Greece and Rome. What she saw on her first visit to the Pan-American on April 28th, three weeks before the official opening, was nothing of the sort. Arriving at the site north of the city, and ducking under scaffolding and skirting piles of bricks, she discovered a kingdom of domes and spires, ornate porticos and arcades. Planned by a group headed by New York architect John Carrère, the grounds next to Delaware Park took up a mile north to south and half a mile wide—350 acres in all—and were laid out in an inverted "T." The long Court of Fountains made up the shaft, the wide Esplanade served as the base. The buildings were covered with the same plaster concoction, called "staff," that had adorned Chicago's edifices. This romantic fantasyland, though, was a far cry from Chicago's scheme. Buffalo's designers, hoping to compliment their friends in Latin America, had wanted an "American" style, and they had found one, sort of, in the mission buildings of Spanish America—in Mexico City and in California. "It is in a sense indigenous," explained one of Mabel's guidebooks, for it "symbolizes the European conquest of the greater part of the Western Hemisphere." Mabel didn't note the irony. Wasn't Spain the colonial power that had just been driven out of the hemisphere?

It wasn't just the turrets and towers that amazed Mabel. It was the Exposition's color—its red-tiled roofs, gold minarets, blue domes, and

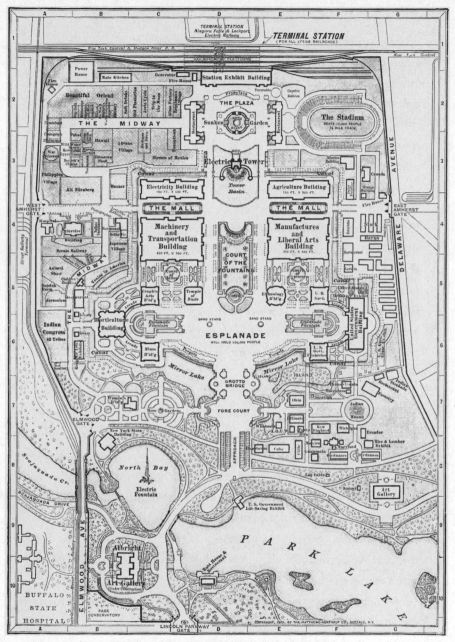

Map of the Pan-American Grounds, 1901. The Albright Art Gallery, lower left, was not completed in time for the Exposition.

yellow, orange, and red tinted walls. It had a hint of green, too, in every big façade, a hued homage to Niagara Falls.

The colors of the exposition were meant to do more than simply delight visitors, Mabel noted. Drawing on the same theme that inspired the five hundred sculptures that dotted the fairgrounds, muralist Charles Yardley Turner used color to symbolize the struggle with the elements and "the progress of the race." The "savage races," he asserted, "have manifested a fondness for strong colors," while "civilized man prefers greyer notes." He painted the bodies of the Horticulture and Ethnology Buildings, near the southern entrance, shades of orange; the Temple of Music was "pure" red. He wanted fairgoers to be soothed by a whiter, more muted palette as they moved northward. The Machinery and Transportation and the Electricity Buildings all sported exteriors of light or dull greens, yellows, and grays. The focus of the fairgrounds, the four-hundred-foot-tall Electric Tower, the proclaimed acme of civilization, was tinted in the lightest ivory, cream, and gold.[11]

The Electric Tower thrilled Mabel. It was, she wrote, "the highest and finest note of all." The architect, New Yorker John G. Howard, would have been pleased. He had studded the small skyscraper with thousands of lights, and, inside the grillwork, placed an elevator that rose and fell like a moving jewel. At the very top of the building, he had planted a gilded sculpture, the Goddess of Light, symbolizing the triumphant harnessing of the natural world. Visitors like Mabel probably hoped the goddess was harnessed herself, for, made of zinc and iron and weighing more than one and a half tons, she would fracture more than herself in a tumble.

Mabel was too proper a Victorian to notice, or too proper to say she noticed, the form of the tower. Other observers were not so shy. The tall, steel-ribbed structure that dominated the grounds, and the waterfall that gushed from it, were suggestive. As one fairgoer put it, "The shaft of the electric tower . . . makes a tender nuptial with the sky and seems to palpitate with beautiful life." The tower's designer referred to it as "the triumph of man's achievement."

If Mabel entertained such notions about the fair's magnificent erection, she never let on. She was nothing but circumspect, in fact, as she toured the grounds and marveled at the grand exhibit halls, with names familiar by now to the fairgoing public: Manufactures and Liberal Arts, Mines, Agriculture, and Electricity. She took notes on the big guns, bulky machines, and mighty engines. She weighed herself on a gigantic set of scales and "tipped the balance" at 119¼ pounds. She also toured state and country exhibits, and reveled in the fair's oddities: a revolving globe made entirely of seeds, a big bear made out of raisins, an elephant constructed of nuts, and a California mission sporting lemon and orange walls. Sculptors had been hard at work, too, crafting a triumphal arch from soap, and the Minnesota capitol building out of 1,600 pounds of chilled butter. Louisiana exhibitors had restrained themselves—they left one of their star attractions, a 122-pound sweet potato, intact.

Being of modest means, Mabel Barnes and Abby Hale brought box lunches and sat to eat on benches or grassy banks. Likely they turned their backs to the crowds—respectable women usually found it embarrassing to eat in public. At other times, the two women sampled the free offerings in the Manufactures and Liberal Arts Building. They tasted "some mighty good doughnuts fried in 'Cottolene'"— beef tallow and cottonseed oil—and tried a tablet of coffee.[12]

It wasn't until July that Mabel stayed late enough to see the biggest sensation of the fair—the Illumination. Just as the colors of the city began to fade, Mabel made her way to the Triumphal Bridge, at the southern entrance. Nearby, in the Esplanade, a band trumpeted a few patriotic airs and then stopped. What took place next, Mabel said, happened before she realized it. Pinpricks of red—filaments in thousands and thousands of tiny incandescent bulbs—began to outline the minarets and domes and arches. The Electric Tower itself began to glow, the teacher said, "like the first flush which a church spire catches from the dawn." The light expanded until it outlined every edge with a daylight yellow.

Electrical engineer Henry Rustin had produced this magic by reducing the unit of light to eight candlepower. Lights at previous expositions had been concentrated and "blinded the eye." Thanks to the marvels of a gigantic rheostat, though, the Pan-American's lighting seemed to echo the slow progress of the sun. Mabel didn't acknowledge the cynics who said the Illumination was a technical necessity— a sudden flip of a switch and a surge of voltage, they declared, would burst the tiny bulbs. She was simply happy to revel in the way that modern science had beaten back the night.[13]

The East Esplanade at night. A bandstand attracts crowds in the foreground; the Ethnology Building stands behind it.

V
WHAT MABEL MISSED

Mabel Barnes was a dedicated teacher, and a committed student of the Pan-American Exposition. But she was also a young woman, closer to girlhood than middle age, which is likely why, out of all the sights that pleased her, none matched the Midway. She loved the disorder: the howls and shrieks, the cymbals and brass, the fun houses, the rides, the animals, the mix of people. The Midway offered the same commentary on race and civilization as the formal fair, but instead of glass cases and exhibit stands, it presented human acts. Advertised as silly, strange, and outrageous, the Midway featured Mexicans, Africans, Hawaiians, Japanese, and Filipinos. Indians took part in a government-sponsored Congress of forty-five tribes; African Americans, in an "Old Plantation," performed the "good old days" before the Civil War.

Chicago's White City had sported America's first midway. The Midway Plaisance, as it was called, received mixed reviews—it was too raucous, too lurid, with too much flesh. People were happily aghast. Fairgoers, it seems, wanted to gape and be titillated. Buffalo officials tried to say no to a midway like Chicago's—they didn't want a "naughty" concourse where women blushed and children grew curious.

Ultimately, though, the dancing girls came: the hootchee-kootchees, the señoritas from Mexico, the hula-hula women of Hawaii, and the Maori "hip-wrigglers."

Buffalo's Exposition Company sold a million dollars' worth of concessions on the mile-long avenue of amusements. They didn't provide an engineering trick like Chicago's Ferris Wheel, but they did offer a Captive Balloon ride and an Aeriocycle, a giant seesaw that sent thrill-seekers 275 feet up in the air. Fair directors tried to distinguish themselves from Chicago with a new name for the concourse, con-

The North Midway. The Aeriocycle (still under
construction) can be seen on the right.

sidering "Queerway," "Climax Way," and "Fort Freak." Ultimately,
they kept things simple. Some people referred to the entertainment
area as the Lane of Laughter, but most just called it "The Midway."

As big a fan as she was, Mabel Barnes didn't like all of the acts on the
Midway. She thought the Streets of Venice were "ramshackle," and she
was "startled" by some of the dancing women. But she couldn't give it
up. At the end of a fairgoing day, hot and nearly worn out with walking, the schoolteacher would enter the wide amusement lane and "fly
the goose" once again. By nighttime, the concourse was littered with
orange peels and popcorn, the air smelled like stale beer, and empty

lunchboxes lay scattered everywhere, like small, foundered fleets. None of this bothered Mabel. She took it all in, and breathed deeply.[14]

The Buffalo teacher missed a few things on the Midway, though. Once in a while she saw actors slip out of character, but she was unaware of other, bigger pretenders. She saw nothing odd about "Wenona the Sioux maiden," the expert sharpshooter of the Indian Congress. Also known as an Indian princess, Wenona was a favorite guest of Buffalo's society women. They didn't realize that thirty-year-old Wenona had actually been born Lillian Frances Smith to a white couple from Massachusetts. Mabel Barnes was also captivated by the "little" Filipinos and loved riding their water buffalos. It was left to others to point out that two musicians in the village were undercover soldiers, planning ways to fight for their country against the United States.

Mabel Barnes missed a lot of things at Bostock's animal show, too, although she could hardly be faulted—almost no one knew. Mabel saw elephant, lion, and jaguar acts; watched a snake charmer; and saw a clown perform with trick dogs. She saw nothing that disturbed her there. It was a different fairgoer who noticed a lion's blood-stained fur, and heard yelps of pain. And at the end of June, on her third visit to the fair, the teacher paid fifteen cents to enter Chiquita's reception room, where she watched the sweet-faced, dark-eyed performer play a miniature piano. She then met the Cuban Doll in person, and shook her hand. Mabel was enchanted and took home the performer's calling card. She had no clue that the Exposition star was actually Mexican. And she had no idea what Chiquita endured, behind the scenes and after hours, at the hands of Frank Bostock.[15]

2.

Summer in the City

I
THE MENAGERIST

Just thirty-five years old in 1901, Frank Bostock looked the part of a ringmaster. Often in uniform, he stood six feet tall. His eyes sat small and high in his white face and he wore his mustache in a wire-like swirl. In 1899, when he had applied in person to bring a wild-animal show to Rainbow City, Bostock touted his experience and made big promises. He boasted about his background, which included growing up in a family of animal trainers in England and marrying into the Wombwell family, owners of a world-famous zoo. He had successfully shown his trained animals in Europe, he said, and had staged several exhibits in the United States. Now, he told officials, he would bring a veritable jungle to the Pan-American Exposition. On a large plot of land, he would plant trees and bushes and bring up to 150 lions to "roam around it at their will."

As intrigued as they might have been, Buffalo directors told him to wait. Carl Hagenbeck of Germany, the world's best-known animal trainer, also wanted a place on the Pan-American Midway, and they could offer only one contract. Some directors thought that the experienced German would do a better show.[1]

Bostock needed the job—badly. He had had success with his animal zoos in the United States but hadn't broken through to the

big time. And it is not clear whether he could go back to Britain. His
brother ran a major animal show in Scotland and would compete with
him. Furthermore, he had left London under a shadow. In the early
winter of 1893, British papers reported that Bostock's wife, Susannah,
had charged him with theft and cruelty. She alleged that he had sometimes left her and their children with no food, and that he had struck
her. To top things off, she said, he had recently abandoned the family,
departing for Europe with the eighteen-year-old daughter of a friend.

Bostock answered the charges. He admitted he had left with the
girl, and that he had taken some jewelry and silverplate belonging to

Frank Charles Bostock.

his wife, but "afterwards he had sorted them out." As for the hitting, he had indeed struck his wife, "because she was drunk, but he was extremely sorry and made it up with her." When Susannah confessed to drinking, the charges against the menagerist were dropped.[2]

After reuniting with his wife and children, Bostock traveled to the United States and started business anew. Stories of his domestic problems did not travel with him; nor did the charges of animal cruelty that occasionally appeared in British police records. Instead, the showman disseminated press accounts that stressed his kindness. There wasn't even much of a fuss when, in the spring of 1900, a reporter who went to see Bostock's lion school in Baltimore discovered signs of rough treatment. He noted that it seemed "necessary to lash [the young lions] severely," and that Bostock's trainers used a long whip with "stings." The reporter also saw that obstinate cubs went without food if they balked at posing for tableaux, and they were only willing to jump through a hoop of fire if they were first cornered, then frightened, by the sound of a gun.

Even if they had read this account, members of the public might not have been bothered. Such training might have seemed more like discipline than cruelty. In Buffalo, the one person who might have been alarmed was Buffalo's mayor, Conrad Diehl. Throughout his public life, the physician not only had championed the needs of sick and poor people but also called attention to the suffering of animals, particularly city dogs. Mayor Diehl, however, had little to do with the hiring of Midway acts.

The choice of a wild-animal show for the Pan-American fair dragged on. Through the summer and fall of 1900, Director of Concessions Frederick Taylor held meetings about Hagenbeck and Bostock behind closed doors, and he refused to comment. Meanwhile, Bostock pumped up the charm. He reappeared in the Queen City twice in 1900, bringing with him enticements such as Chiquita and suggesting new performances. He proposed building a spectacular circular lion cage and acting out "Daniel in the Lion's Den." He himself would play Daniel. The Pan-American Exposition, he predicted, would be "the largest ever held in Europe or America."[3]

After he got the job, Bostock applied the same marketing skills to his show that he did to selling himself. His wild-animal arena was one of the first concessions to open in Buffalo in May, and his press agent announced it as the "star feature" of the Exposition. Mirroring the themes of human conquest that the fair trumpeted elsewhere on the grounds, the agent claimed that the Animal King ruled over a vast domain. He had conquered the natural world to assemble his collection, having invaded the "jungles of Africa, the forests of India and the deserts of Arabia and Egypt." He had gone on to "ransack" ice floes and frozen seas. Not surprisingly, he not only had triumphed over other "animal subjugators," but he also had overpowered indigenous peoples wherever he went: "The savage natives of the globe's remotest corners and most inaccessible recesses know of the white man who rules the most ferocious beasts into absolute submission and obedience by some mystic power they cannot understand."

Bostock's trained animals seemed the perfect match for the Exposition's pronouncements about race. One of Bostock's monkeys was linked to American foreign policy in Southeast Asia. Mike, a small orangutan from the Philippines, couldn't comment on the war in the Philippines, noted a bystander, but "he is of peculiar interest because, in some respects, he resembles the creatures we are daily filling with Maxim bullets and shells. . . ." The showman's celebrated chimpanzee, Esau, was alleged to be "the missing link" and a member of a group of ape-men recently discovered in Uganda.

Press offices admitted that Bostock's charming publicity agent, Captain Jack Maitland, had them cowed. Maitland's releases, one reporter admitted, were often sent directly to the composing room. "It doesn't make a bit of difference what the press agent writes about or how much truth there is in it," he said, "so long as the matter is published and the name of his employer or his show appears." When business becomes dull, he added, "it is positively wonderful to see how the captain's imagination comes to the rescue. The next day the papers contain a startling story of a fight between lions or an animal getting loose with the narrow escape of the employees. An inquisitive public flocks to see the ferocious beasts and the victory of the captain is complete."

Bostock added to his mystique by staying behind the scenes. When he did make an appearance, it was rarely without an animal on his lap, and usually in the company of his bodyguard, a six-foot-tall former major in the Bombay infantry. Turbaned, and of a "mighty athletic build," the guard had reportedly hunted and killed up to a hundred tigers.[4]

As popular as his show proved to be in Rainbow City, the Animal

Animals line up in front of Bostock's arena. The large elephant in the center is likely Big Liz, who befriended Jumbo II.

King had his troubles. At the end of June, a Bostock trainer, Herman Weeden, suffered a nervous breakdown. Weeden had returned to work after a tiger attack—too soon, it turned out. He began hallucinating. One evening he imagined that he was Bostock himself, and, wielding an imaginary whip, flew around his quarters in a frenzy, trying to subdue phantom animals. "Look," he shouted, "I am the animal king. I am the monarch of the brute world." Bostock, endeavoring to placate Weeden until an ambulance arrived, played the part of a lion, crouching on all fours.

Although Frank Bostock could manage the temporary loss of one of his trainers, he wasn't so sure whether he could get along without his Cuban Doll. Sometime in late June, one of his employees reported to him that Chiquita had a growing attachment to Tony Woeckener, a young cornet player in her show. Bostock was enraged. Chiquita, he felt, owed him unwavering loyalty. He had befriended her, supported her, managed her, taken her into his home. He had provided her with jewels, dresses, furs, tiny carriages with tiny horses, and even a car. He had made a lot of money with her, to be sure. But how dare she? The Animal King was nothing if not resourceful. It would take only a few small lies and, if that didn't work, a lock and key.[5]

II
THE CUBAN ATOM

The journey that had brought Chiquita to the Pan-American Exposition and placed her under the strong arm of Frank Bostock spanned at least three countries. Before she became the Doll Lady, or the Cuban Doll, or even Chiquita, she was Espiridiona Cenda, born in Guadalajara, Mexico, around 1878. We do not know whether her birth was attended by any cries of surprise, but it was reported that when she entered the world, she was eight inches long, three inches wide, and, when doubled up, the size of "a large grape." More impressive than her size, apparently, was her proportionality. "Nearly all midgets have some deformity," the Buffalo press would explain, "something freakish in their appearance, but this little woman is one of the very rare exceptions." As her managers would claim, she was "a perfect living, breathing doll."

Alice Cenda, as she was known offstage, attended school in Mexico and trained in singing and dancing. When she was about fifteen, her father hired a manager to exhibit her to the public, and she toured widely in the country, and later throughout Cuba. Sensing profits in using her as an emblem of the island, one of her managers reinvented her as Cuban and brought her to New York. Two brothers came with her.

In New York, in 1895, Frank Bostock—who later would say he found her as a "Mexican peasant"—took over her contract. He paid her $80 per month for her performances (and held onto her money) and brought in, by one account, $1,000 per week. By 1900, she had become one of Bostock's main attractions and toured in the United States and Europe. Like so many other Little People in show business at the time, she met heads of state and royalty.[6]

In establishing Chiquita as one of his big acts, Bostock joined the swelling numbers of American freak-show promoters. Human "oddities" had been put on display for generations, but they reached their peak in the United States at the turn of the twentieth century, with the advent of shows like P. T. Barnum's, and with touring circuses and fairs. Individuals with extraordinary bodies were fascinating to a culture obsessed with classifying, measuring, and establishing standards and norms. So much appeared in flux in American society at the time that spectators looked to disabled people for reassurance that there were limits to fluidity. People such as beleaguered urban laborers or immigrants or the rural poor—those who commonly attended the shows—were reassured that they had one attribute—the shape or size

Alice Cenda, aka Chiquita, in the late 1890s.

of their bodies—that fit an ideal. The most popular acts were those like Chiquita's, which featured people of non-European ancestry.

Buffalo audiences took to the Doll Lady with enthusiasm. They enjoyed her singing and dancing, her "quaint" accent, and her stories about meeting monarchs such as Queen Victoria. Like Mabel Barnes, they also enjoyed shaking her tiny hand. Not only did Chiquita sell a lot of tickets, but, on July 10, Director-General William Buchanan christened her "The Mascot of the Pan-American Exposition." It seemed a fitting title. The little "Cuban" was a perfect symbol for what the United States sought in its relations with Latin America—and with the island nation in particular. During and after the war with Spain, the American popular press had pictured Cuba as a charming daughter to Uncle Sam, eager for his aid and protective oversight. Chiquita, who exhibited "the charm of manner so characteristic of her race," and who could "sit with ease and comfort in the hand of any grown adult," seemed the perfect embodiment of this thinking.

Chiquita performed her part perfectly. She spoke out about the Cuban fight for liberation, and was given special authority to do so. As early as 1896, the press announced that she had been born into an established family in Cuba's Matanzas province and that Spaniards had torched her family's home and sugar plantations, and put her to flight. With two of her brothers in the army of *Cuba libre*, she was "a little rebel every inch of her 26 inches and she hates everything Spanish with the hatred of a loyal daughter of Cuba."

During the winter of 1901, Chiquita had even become friends with the chief architect of America's Cuba plan, William McKinley. On February 13, dressed in diamonds, silk, and satin, she had been carried up the White House steps into McKinley's office. "I want to thank you," she told the president, "for all that you have done for my people." McKinley, delighted by her speech, unfastened a pink carnation from his lapel and pinned it to her dress.

Chiquita may have personified the hopes of American policymakers, but she was neither as compliant nor as needy as her size suggested. Nor was she as dainty as her act implied. While she had domestic talents, especially in embroidery, she was also a spitfire of spunk and nerve. She played a hard hand of poker. She put up "a pretty stiff little bluff on even a 'bob-tail flush,'" claimed one reporter. And he suggested that it wasn't just the card game. She didn't want to be treated like a toy in any setting. "Perhaps nothing so well illustrates her dislike for being regarded as different from other people," he commented, "as the fact that she objects to a high chair at the table even though her little chin comes only to the edge of her plate . . . but she suffers

Bostock's Main Attractions.

the consequent inconvenience to herself because of her pride . . . she wants it understood that she is not a doll."

But Bostock liked her as a doll. Even though the showman was just over a decade older than Chiquita, she called him "Papa." While on tour, he gave her sleeping quarters with his children and encouraged her to take meals with them. He sometimes seated her with a different dining companion, too—Esau, the trained chimpanzee. Esau wore human clothes, slept in a bed, walked erect, and played the piano. He was also "learning to talk." Bostock liked partnering brown-skinned Chiquita with the monkey so much that he double-billed them as the two main sideshows of his zoo.[7]

Chiquita's friend Tony Woeckener was, by all accounts, small for his age. Not Chiquita small, but short and slight enough that people called him "Little Tony" and said he looked at least two years younger than he was, which was seventeen. The earnest, open-faced teenager, who kept his hair in a determined part, played the cornet outside Chiquita's reception hall, helping to lure fairgoers into her show. He had first met Alice, as he called her, in 1900, in Atlantic City, when he and his family—seven musicians in all—had been hired by one of Bostock's shows. "She looked at me and smiled, and I smiled back," Tony remembered later. "She heard I sang, and . . . I did sing, and that was the beginning of the love affair."

At Buffalo they grew closer. "About two weeks after the big show began," Tony recalled, "I was sitting with her in her dressing room. A pretty warm friendship had sprung up between us. . . ." He decided to ask her to be his sweetheart. Alice said yes.

They tried to keep their affection for each other a secret—from Bostock, from the other showmen, and from the public. Tony surreptitiously brought over dinner after her show ended for the day, and they spent time together, quietly. "I taught her to write English," said Tony. "She taught me some Mexican."

They knew they risked discovery. They were well aware that when the Animal King found out about their liaisons, they would suffer. And sure enough, one day at the end of June, one of Bostock's men witnessed something, or heard some gossip, and carried the news to the boss.

Frank Bostock didn't just fire Tony; he let a rumor circulate that Tony had almost destroyed Chiquita's whole concession. The musician, he claimed, had broken an expensive calcium light and nearly started a fire. Tony, he clearly hoped, would have a hard time finding a new job.[8]

III
THE VANISHING STATE

By July, most people in the city of Buffalo had been swept into the whirl of the Exposition. Even those far out of town did not escape. The Electric Tower sent its searchlight across the dark river to Canada, and the percussion of fireworks carried so far through the humid air that even farmers in the hinterland felt the thumping celebration.

Word of the fair's magical lights, colors, and Midway was also carried across the country, in nearly every newspaper. It made its way into the *New York World*, and, from there, into the hands of the *World*'s curious readers. One of them sat in a rocking chair in a small house in a Michigan town on Lake Huron. Down on her luck and discouraged, Annie Edson Taylor read about the crowds gathering that summer in Buffalo and in nearby Niagara Falls. Suddenly—she felt nearly blinded from the shock—she was struck by one of the strangest ideas imaginable. She had thought of a miraculous way to pay her bills. She would take advantage of the throngs and do something so startling and extraordinary that fame, fortune, and a happy future would be assured.

Sometime later, and quietly—she did not want to share her plans—she gathered cardboard, cut it in strips, and sewed the strips together. Then she stepped inside. She had designed herself a barrel. What she would do with the barrel was drop herself 160 feet over Niagara Falls. If she lived to tell the tale, she would become an Exposition sensation, perhaps even a sensation for life. No one had ever survived such a descent.[9]

Even as Annie Taylor learned of the Pan-American's big crowds, Buffalo itself began to fret. World's fairs were put together by individuals —businessmen, publishers, politicians, clubwomen—but quickly they became a referendum on a place, a place that, as time went by, seemed increasingly human. Buffalo struggled for recognition, approval, and a public pat on the back. And it sought to make a lot of money. As June days edged into midsummer, fair investors and directors infused their pride with worry and began to take measurements. How many clicks had the turnstiles recorded? Of the nearly forty million people who lived within a day's journey, how many were making the trip to the city? Who had written glowing words about the show? Who had been critical? Ticket counters realized that Omaha's numbers were being surpassed at the Pan-American, but not easily. Chicago's exposition, that king of fairs, seemed to be pulling out of reach.

Local editors picked up the concern and tried to help. They printed every scrap of praise, no matter how lukewarm or hyperbolic. In the middle of July, they announced that construction of the fair was finally finished. They claimed that Rainbow City was cleaner and neater than other fairs. (It featured asphalt, of course, so there was *very* little dust.) And they boasted that while Pan-American visitors enjoyed breezes off the lake, New York City was "sweltering, Philadelphia lies languid . . . and Chicago's list of heat prostrations grows."

Exposition officials did their part. They cut the price of admission on Sundays from fifty cents to twenty-five cents to attract working families. And they looked to special "days," when the fair honored particular groups or states or countries, for increased attendance. Sometimes they assigned three groups to one day to triple the chances for a big turnout. July 24, for instance, served as YMCA Day, Utah Day, and Knights of Columbus Day.

This strategy worked—sometimes. Utah Day arrived, but Utah, it seems, couldn't make it. Visitors to the Exposition looked in vain for displays of Utah's produce, manufactures, and art and literature. They looked for the Utah Building and for people from Utah. Not surprisingly, they looked for Mormons with multiple wives. And they looked for the man who had proposed Utah Day in the first place. Apparently, even he had left town. One reporter described a frustrated visitor who had once lived in Utah, circling the entire Exposition for signs of his former state. Footsore, he sat down on the steps of the Ohio Building as the sun began to set. "Where in the name of Christendom is Utah?" he muttered.

Slow days were indeed a puzzle. Why were the numbers off? Was it the fault of the railroads? Were the fares off-putting? Was the advertising weak?

People wondered whether it was a mistake not to have had a special place where all women were welcome to gather. The Women's Building, a veranda-wrapped house at the southern entrance of the grounds, was intended mostly for private use. The headquarters of the Exposition's women managers, it held meeting rooms for clubwomen, college graduates, and women of other "leading" organizations. There were no "women's" exhibits in the building, and not many more in the rest of the fair. This, the Board of Women Managers insisted, signaled progress.

Chicago, some murmured, had done things differently.[10]

Fair directors also wondered to themselves whether the theme of the fair had anything to do with slow attendance. Perhaps the focus on the Southern Hemisphere had been a risky decision. Middle-class Americans were unabashedly dazzled by Paris, London, Vienna, and Venice. In school, many of them learned the histories and languages of Europe and, if they had enough money, they toured Europe. At expositions, they admired European exhibits. In Chicago's White City, they paid to see French pavilions, German imperial treasures, and Elizabethan architecture. Director-General Buchanan remarked that limiting displays of industry and manufacturing to the Americas had perhaps "weakened to a degree general interest in the exhibits."

Exposition planners had known from the beginning that many of their potential guests held unflattering and poorly informed opinions of Latin American republics. "It is astonishing to learn," commented the Exposition's director of publicity in 1899, "that a very great percentage of the better educated classes are almost entirely ignorant of the wealth of the [South American] countries." A prominent businessman he knew confessed that he associated the name Buenos Aires with images of "dusky natives clad in jaguar hides." The publicist pointed out to him that the cosmopolitan Argentine capital had plenty in common with Paris.

The press certainly played a part in assigning second-class status to Latin America at the Exposition. It billed New World exhibits as exotic and romantic, and let loose a barrage of backhanded compliments. Argentina demonstrated "marvelous growth and progress." Chile, which had spent $500,000 on its Buffalo building, was described as "progressive and highly advanced" and was commended for having sent its young artists to Europe for training. Then there was Mexico, which showed "wonderful advancement," and Cuba, the "little American isle," ripe for investment.[11]

IV

MIDWAY DAY

Whatever the reason for slow ticket sales, Midway men were ready to tackle the problem. They knew that visitors to Rainbow City sometimes bypassed big exhibits on machines and art and history but rarely, if ever, missed the Lane of Laughter. The director-general might make all the noise he could about edifying the masses with grand buildings, but they knew—Chicago had taught them this—that a midway could make or break a big show. What the Exposition needed was a Midway Day. Let the misfits, the animals, the clowns, the "foreign" people take over the whole show. "You've got to be tawdry," said one concession owner, when asked how to sell tickets.

Exposition officials agreed, and on August first they sent marketing men to the railroad offices to get special rates. They hired a hundred girls to accost strange men on city streets and pin satin ribbons to their lapels: "Midway Day, August 3d, Pan American."[12]

Two days later, the sun shone, the winds off the lake blew cool, and the crowds materialized. Pennsylvania and Ohio trains had to order extra coach cars, Rochester's platform was packed with fairgoers, and, on Buffalo streetcars, people sat on laps and stepped on toes. On the grounds, animal trainers rose early to wash elephants and grease camels, and performers dressed themselves in furs, feathers, grass skirts, and paint.

It was late morning when the parade began. Like a serpent nosing into forbidden territory, Midway floats edged into the Esplanade, then onto the Main Court, where visitors waited, fifteen to twenty deep.

Members of the Carlisle Indian band, with their bright red uniforms and brass instruments, led the way, along with concession managers. Calamity Jane and Geronimo followed, then a band of Apache, and other warriors on foot and on horseback. When Calamity Jane's mules were jostled by an Indian chief, a spectator shouted: "Teach him to talk English, old girl!"

Chiquita, in a yellow brougham pulled by a miniature pony, led the Bostock concession. The Animal King, pushed by an attendant, sat in a bicycle chair with a tiger cub on his lap. Behind him, his bodyguard, the Indian, followed closely. The elephant Big Liz carried adults on her back, and smaller elephants bore children. A polar bear padded by, then a grizzly bear.

The crowd sent up a roar at the sight of the string band from the Filipino Village. The Igorrotes rode water buffalo and towed their children in oxcarts. The audience whooped its approval, too, at the armed Moros and Tagalog soldiers, the "little fighters" who had just helped oust Spain from the Philippines. One observer noted that these men could certainly use the military know-how of the United States. The Tagalog fighters were "armed with old fowling pieces, blunderbusses and muskets which looked as though they had been handed down from father to son for generations."

Of all the floats that wheeled by, none provoked as much glee as the Old Plantation's watermelon. Fifty feet long and fifteen feet wide, the watermelon had a rind as "green as the sea and its core [was] as red as a bullock's heart." One witness could barely contain himself: "The seeds—prepare to laugh—the seeds were the heads of live darkies sticking out with a grin on as broad as the sunset over Tierra del

The Indian Congress parades along the Court of Fountains.
The Electric Tower rises in the background.

Fuego. Yah, yah, yah, they laughed and their teeth shone and their eyes glistened." Behind the float, one hundred other Old Plantation actors performed the cakewalk. "All of them," said the spectator, were "as gorgeous and gay as a stalk of hybrid hollyhocks."

A band struck up "Dixie." The crowd gave a cheer and joined in, swaying to the song.

The parade wound around the Court of Fountains and reentered the Midway. But before it did, an enormous elephant reached his trunk out to the crowd and plucked a hat from the head of a boy. With a cannonlike snort, he fired it in the direction of the Electric Tower. The crowd howled.[13]

Jumbo II had arrived in Buffalo in the third week of July. Shipped in a box with his head barely visible, the elephant had been pulled uptown by thirty horses and rolled into the Exposition grounds with fanfare. Jumbo carried a romantic history—Bostock made sure of that, and he and Captain Maitland had been bragging about it for days. Although some people might have suspected—especially if they read New York City papers—that this was the same Jumbo II that had been captured as a baby in India, raised in Germany, and unloaded onto Brooklyn docks in 1900, they were informed otherwise. This elephant, said to be twelve feet tall with a weight of nine tons, had a much more compelling résumé. Once known as Rustin Singh, he had been the valued possession of an Indian prince. As a gesture of goodwill, the prince offered him to the Duke of Edinburgh, but he was too big to travel to England. Instead, he served in the British-Abyssinian campaign, and, along with nineteen other elephants, hauled mountain batteries to the front for Lord Roberts and Charles Merewether. At the Battle of Magdala, he was wounded but labored on, and when other elephants threatened to break rank, his trumpeting kept the herd together. For his valor, Rustin Singh had been decorated by Queen Victoria and "pensioned for life."

Jumbo II arrives in Buffalo.

By some miracle, the same animal now stood before the admiring crowd in Buffalo. Bostock claimed he had staged a coup. London had wanted the elephant, he said, and members of Parliament had been shocked that the "historic beast" had slipped away. The director of Regent's Park Zoo, responding to London's indignation, had just cabled Bostock to name a price to have him sent back to England.

Bostock held onto his prize. He renamed the elephant Jumbo II—

commemorating the world-famous Jumbo who had died a decade and a half earlier—and brought him to Buffalo. The new elephant had a rough time of it at first. His stall wasn't big enough, and he seemed to be lonely. He thrashed in his pen. He also seemed more timid than expected; he was afraid of thunder. But after a few weeks he settled down and made friends—close friends—with Bostock's other large pachyderm, Big Liz. They had begun to trumpet to each other. And today, on Midway Day, he was the life of the party.

Among the floats in the Midway parade as it moved into and out of the Court of Fountains, was Lubin's Picture Machines. Spectators were asked to stay still as Lubin's big camera clicked past them, shuttering open and closed. Mabel Barnes was probably photographed. She had met friends and cousins at the East Amherst gate, and, after navigating through the thick swarms, had watched the parade from the Tower Bridge. Fred Nieman may have been captured on film, too. The family he lived with said he came into Buffalo like clockwork. They didn't say whether or not he took his revolver with him. He kept one, apparently, in his bedroom.[14]

3.

The Favored Guest

I
CANTON

Midway Day gave the fair's directors new hope. The Pan-American had drawn the "biggest crowd of the Exposition": more than 106,000 visitors. The next day, however, the crowds were down again, and exhibitors wrung their hands. The Exposition had sold fewer than three million tickets, and it needed to sell three times that many to make ends meet. It was, as one reporter put it, "high noon" in Rainbow City.

Officials remained calm. Chicago had been slow to take off, they said. Director-General Buchanan offered soothing words, stating that most people liked to travel in the fall, when it was cooler.

Press agents concurred. Just look at the weeks ahead, they argued. That swashbuckling hero of expositions, Buffalo Bill Cody, was on his way. Always attuned to the very latest in global skirmishes, Cody had added several new shows. He had retired the "Charge Up San Juan Hill" reenactment and replaced it with a scene from the Boxer Rebellion in China. His soldiers would perform "The Taking of Tientsin."[1]

But better even than Buffalo Bill would be the most illustrious guests ever: the president and the first lady of the United States. It did, however, seem wise to verify the arrangement, so, less than a week after Midway Day, Mayor Diehl, Exposition President Milburn, and

Director-General Buchanan boarded a train for Canton, Ohio, site of the summer White House.

The men made an odd trio. William Buchanan, a stout forty-eight-year-old with a stiff, brushlike mustache, had made a name for himself managing Corn Palace Festivals in Sioux City, Iowa. His oversight of these popular oddities, which featured a palace made entirely of husks and grain, had qualified him to supervise three departments at the Chicago World's Fair in 1893. He had done such a good job that Daniel Burnham, the White City's director of works, had given him a recommendation for Buffalo. In addition, a diplomatic post as minister to Argentina, where he learned Spanish and studied the "character of the Latin American," provided him with the international expertise that seemed perfect for the Pan-American. He was also respected for his Latin American stock portfolio.

Pan-American Power Brokers. Buffalo Mayor Conrad Diehl (*left*); Exposition President John Milburn (*center*); Director-General William Buchanan (*right*).

Of the three men, John Milburn was most like McKinley himself. A well-fed, curly-haired man, six feet tall, Milburn had grown up in Sunderland, England, but in his late teens moved to be near relatives in Batavia, New York. There he trained for the bar. Relocating to the big city of Buffalo, he became one of the most successful corporate lawyers in the state. Known for his velvety eloquence, he defended the "rights of capitalists." Success begat success, and he soon surrounded himself with people of wealth and position. A friend of President Grover Cleveland, he became a "business-friendly" Democrat and a leader in local social clubs. He lived on Delaware Avenue.

Milburn was in league with Buffalo's elite, but he did not immerse himself too deeply in his adopted city. He sent his sons to boarding school in Philadelphia and to college at Oxford, where they played polo. By the time he was tapped to lead the Pan-American Exposition, he was also leading a double life. The Exposition captured half his attention, but a scandalous trial occupied the other half. In one of the country's most sensational murder trials, where wealth seemed pitted against truth, Milburn defended New York society man Roland Molineux. A member of New York City's famous Knickerbocker and New York Athletic Clubs, who was also known as a hot-headed "aesthete," Molineux had been convicted of poisoning a female lover. In June 1901, with the Pan-American in full swing, lawyer Milburn persuaded a panel of judges that Molineux deserved a new trial.

In many ways, John Milburn and Conrad Diehl represented Buffalo's flip sides. The son of German immigrants—his father was a stonemason—the quiet-mannered Diehl had put himself through the local university and medical school. He was a familiar sight to city residents at the turn of the century, trotting along in his buggy with his black dog, Jack, by his side. When he was elected mayor, he served as physician for 5,000 Buffalo families. He also advocated on behalf of the poorhouse, the orphan asylum, and local schools, and he lived in the same small brick house on West Genesee Street where he had been born. "Buffalo is a word that means all to me," the mayor had announced in 1899, at the Exposition's launch, and probably no one wanted the fair to succeed more than he did.

As a Democrat and an advocate for Buffalo's poorer communities, Conrad Diehl had not wanted too many businessmen dominating Exposition planning. Money won out, however, and at board meetings, the balding, soft-eyed Diehl sat across from many of his adversaries, some of whom had vocally opposed his election as mayor. John Milburn, in fact, had labeled Diehl's support for more equitable distribution of wealth and an income tax as "socialistic and anarchistic" and suggested he be "thrashed and thrashed soundly" at the polls.

Conrad Diehl, fifty-eight years old in 1901, was anything but anarchistic. But he believed that government should take a stand against concentrated wealth and monopolies, and he supported William Jennings Bryan's position against the country's aggressive imperialism. To rich men and investors, all this sounded frightening.[2]

On the way to Canton, though, the men traveled with a single purpose, and they had reason to be optimistic. They knew that the president loved fairs and that he would surely want to see this one. It mirrored so much of the America that he cherished. And he had actually been to the fair before, sort of. Back in 1897, when Cayuga Island near Niagara Falls had been proposed as the Exposition site, the president had driven in the very first stake.

The president was a pushover. He and his secretary settled on the first week of September for his visit. It would be a significant occasion for many reasons, but most especially because it was Ida McKinley's first major appearance after her long convalescence. The American people would be pleased to see her on the mend.

II
THE SUMMER WHITE HOUSE

The McKinleys had spent the height of the summer at Canton, gliding quietly from one day to the next. They had entertained a few friends and officials, but by and large their days were marked by simple domesticity. Such a life suited the twenty-fifth president. At fifty-eight, William McKinley was not a vigorous man. He had once had a tougher constitution—it would have been hard to survive the Battle of Antietam without one—but those days were behind him. At slightly more than five and a half feet tall, he carried most of his weight in his chest. His eyebrows gave some alertness to his round face—they had the tilt of a screech owl—but they did not take away from his stolidity. Americans appeared to like this about him. He had built his reputation on his slow and steady ways, as well as on his desire to listen to the will of the people. One politician claimed that his ears were full of grasshoppers, so closely did he keep them to the ground.

It would have been hard to tell from his unruffled and genial demeanor, but McKinley had made controversial decisions since taking office in 1897—such as going to war with Spain, seizing control over Cuba, and then taking the Philippines as an American possession. Democrats, who called him Emperor McKinley, rallied behind the charismatic William Jennings Bryan, who ran against McKinley in 1896 and again in 1900. "Running" was too strong a word for the way McKinley himself campaigned. Avoiding debates with the masterful Bryan, and wanting to stay close to his wife, he had remained in Canton in 1900 and speechified from his veranda. While Bryan rode the rails around the country, reddening and sweating with political bluster, McKinley literally stood his ground.[3]

Now, six months into his second term as president of the United States, he was at ease with his job. The American economy was continuing its steady rebound from the depression of the 1890s; the climate for business and trade—McKinley's pet concerns—was favorable; and foreign crises had subsided. Washington had settled into its seasonal repose. The only thing that seemed to challenge the president was his wife's chronically poor health.

Ida Saxton McKinley was a delicate woman, with an upturned mouth and blue eyes that belied her discomfort. Since the deaths of her daughters, an infant and a four-year-old, in the 1870s, she had suffered from epilepsy, depression, and a need—sometimes a desperate need—

William and Ida McKinley, in foreground, at
a dinner party in Cleveland, ca. 1897.

to be near her husband. He in turn was devoted to her, and to ensuring her dignity. On one legendary occasion, at a luncheon party, he anticipated a seizure and placed a handkerchief over her face to avoid embarrassment. Recently, her health had worsened. Their cross-country train tour in the spring had been interrupted by a near-fatal infection, and the couple had devoted the summer months to her convalescence.[4]

When William McKinley was not dealing with government business, he ambled about Canton, sauntering downtown and walking to church. He preferred to do so without escorts. The press asserted that his willingness to mingle with the public demonstrated his confidence. Unlike kings or tyrants who feared violence at every turn, this elected man trusted his constituents. Dismissing protection was also a sign of strength. Abraham Lincoln had been labeled a coward when he had worn a disguise while traveling through Baltimore in 1861, and he had never heard the end of it. Indeed, the criticism stung so painfully that Lincoln thereafter traveled in open carriages. As the war dragged on and his enemies and critics became increasingly bitter, he was assigned more and more protection, but he continued to object. It is likely he felt liberated when, after Lee's surrender in April 1865, he pared down to a single bodyguard.

William McKinley may have dismissed security officers for another reason. Some members of Congress believed armed guards expanded the power of the executive in unhealthy ways—they implied a standing army. Lincoln's assassination in 1865 and Garfield's in 1881, legislators argued, were the result of unusually fraught moments in history, and they did not signal the need for more security men.

It was during Grover Cleveland's second term in office in the 1890s that sentiment began to shift toward greater protection. Responding to the turbulence and anger generated by the 1893 depression, Cleveland's wife and personal staff encouraged the Secret Service—which was originally formed to thwart counterfeiting—to collaborate with the Washington Metropolitan Police to guard Cleveland. The president's opponents didn't miss a beat. They claimed he was acting like royalty.[5]

Things hadn't changed much by 1901. It was still important for the chief executive to appear stalwart and trusting, and presidential aides and family members continued to fret about safety. Ida McKinley had feared a second term in the White House for the exposure it gave her husband, and she was well aware that the War of 1898 had spawned new enemies. She fretted about his speeches, his trips, any large public gathering. The president's personal secretary—a nervous, bespectacled lawyer named George Cortelyou, who imagined snipers and bomb-wielding intruders around every corner—also kept a file called "Assassination Plots." He filled it with newspaper clippings about threats, predictions, and attempted killings.

Cortelyou and other members of McKinley's staff were particularly worried about anarchists—those who opposed government in all its forms. Anarchists in the United States at the turn of the twentieth century, numbering ten thousand at the most, were mostly centered in urban working-class and European immigrant communities in the Midwest and Northeast. While many of these dissidents advocated a stateless society through nonviolence, a series of events led the media to jump to generalizations. The Chicago Haymarket bombing in 1886, attributed to anarchists, and a series of assassinations in Europe in the 1890s, carried out by or linked to anarchists, reinforced stereotypes of anarchists as violent revolutionaries. Worse, by 1901, radical American anarchists were said to be proliferating. In the summer of 1900, the Italian American silk weaver from New Jersey, Gaetano Bresci, had crossed the Atlantic and shot and killed the Italian king, Umberto I.

Four days before he left for Buffalo, President McKinley entertained George von L. Meyer, ambassador to Italy, and General Arthur MacArthur, military governor of the Philippines, in Canton. The topic of anarchists came up. In conversation before dinner, a member of the party mentioned the murder of the young Italian king. The president, in a breezy mood, dismissed such dangers. Such dissenters needed to be tolerated, he said, and he wasn't worried. When Ambassador Meyer showed him a news article about how some American anarchists hoped for more violent acts, the president did not react. "He saw no cloud," said the ambassador.[6]

And, indeed, it seemed only sunshine floated over Buffalo. The city itself had gone for McKinley in the 1900 election, and the Pan-American Exposition, where the eagerness to see him was rising high, was a McKinley fair. From its big guns to its living Cuban and Filipino exhibits, to its sculptures and bridges and buildings that proclaimed American supremacy, it proudly reflected the president.

As someone like George Cortelyou knew, though, a fair that served as a victory celebration was as likely to draw opponents as supporters. In the recent election, six million Americans had voted against McKinley, many of them against his plans for Philippine annexation. Nor was the war in the islands truly over. Even though revolutionary leader Emilio Aguinaldo had been captured by American forces in March—his surrender had been orchestrated by General MacArthur himself—other Philippine fighters continued to resist American occupation.

One of them, in fact, lived close by. Pablo Arcusa, a Filipino performing on the Pan-American Midway, had been good at fooling people. He had impressed "Pony" Moore, the manager of the Filipino Village, with his eagerness to come to America, and the clarinetist had worked hard and been well liked. The fact that Arcusa received letters from Hong Kong, where leaders of the Philippine insurgency had been exiled, didn't raise any red flags. Arcusa said only that he had friends there.

On August 10, though, Arcusa, who in truth hated "Americanos," received some mail from Hong Kong that prompted him to call a meeting in his Midway hut. He told the men who gathered that the fight in the Philippines was still on, and that they could help free their country from American oppressors. While Emilio Aguinaldo had been captured, General Miguel Malvar was carrying on the fight for liberty. They could help the resistance if they would only return to their country with a gun. Smuggle out a musket, a rifle, or a revolver, and freedom would be that much closer. Arcusa was doing his part. When the Exposition finished, he would travel west, talk to Filipinos up and down the California coast, and try to take a regiment back home.

No sooner had Arcusa unveiled his hopes to his fellow band members than several of them ran to Pony Moore and spilled the story.

Arcusa fled the village and the Exposition. He left behind a few papers, including a military commission signed by Aguinaldo himself.

Fair directors had to have exhaled gratefully when Arcusa ran off. They were assured, though, that other Filipino musicians were perfectly happy with American rule over their homeland—some were even hoping to become American citizens. "None was in sympathy with him," wrote the *Courier*.[7]

McKinley's advisers and security men may or may not have known about Arcusa. They were certainly unaware of the disaffected man living as a lodger in West Seneca. And why would they have known? Fred Nieman led an unremarkable life. He read newspapers and ate by himself. Sometimes he left the boardinghouse, rode inconspicuously into Buffalo on streetcars, and then rode back. In mid-August, he made a small dent in the memory of an exhibitor named Cecil Hooke when he appeared at the Exposition at the Manufactures and Liberal Arts Building. Nieman, wearing a new dark suit, asked Hooke for work, saying he had experience as a carpenter. Hooke didn't figure out that Nieman was lying about his skills until he had been hired. He also thought Nieman had worked for other exhibitors, too, for he had been seen about the grounds. But then, Hooke said, he "disappeared."

One morning late in August, Nieman carried his trunk down the stairs of his boardinghouse, took a long look in the mirror, and announced he was leaving. When his landlord's son asked him where he was going, he threw out a few possibilities: Detroit or Toledo or Cleveland. Or Baltimore or Pittsburgh. Nieman may not have had an exact plan. But he read the newspapers; he knew who was coming to the Exposition. There was an opportunity at hand in Rainbow City, and, after taking care of some business, he would be back.[8]

III

TINY MITE

Of all the showmen at the Exposition, probably no one was as deliri-
ously atwitter over McKinley's upcoming visit as the Animal King.
He knew the value of presidential attention and wanted an audience
with the chief executive. Taking a leopard hide (of uncertain prov-
enance), he artfully lettered an invitation to McKinley on the back
of the skin, photographed it, and made sure Captain Maitland sent
it to the press. He also put his minions to work printing advertisements for his show, and he produced so much material that he com-

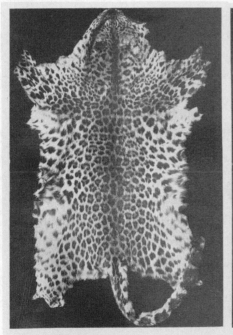

A NOVEL MIDWAY INVITATION TO PRESIDENT McKINLEY
(REVERSE)
THE WORDING IS DONE BY THE BURNT LEATHER PROCESS, NOW SO FASHIONABLE,
ON THE DRESSED SIDE OF A BEAUTIFUL LEOPARD SKIN

A NOVEL MIDWAY INVITATION TO PRESIDENT McKINLEY
(OBVERSE)
BOSTOCK'S LATEST NOVELTY. AN INVITATION TO THE PRESIDENT TO VISIT
THE WILD ANIMAL ARENA UPON THE OCCASION OF HIS TRIP
TO THE EXPOSITION

Frank Bostock's invitation to President McKinley, inked on leopard skin.

mandeered one of his young elephants, Little Doc, to haul it over to the post office.

Bostock also prepared a special act for September, a capstone to his successful season. Trainer Jack Bonavita's command over lions was so convincing that six happy couples planned to tie the knot inside a cage in the company of the beasts. Confident that the McKinleys would be intrigued by the event, Bostock planned to stage a rehearsal of the nuptials, including a display of the golden wedding cage, while the president was in town.

The lion wedding, if it worked, would give the Animal King a boost in public relations. While his concession was one of the busiest and most popular acts on the Midway—often playing to standing-room audiences—the show had suffered a small setback in early August. The incident, which had taken place just after Midway Day, had involved a baby marmoset and Jumbo II.

The baby monkey, Prince Tiny Mite, orphaned shortly after birth, had quickly become a Midway sensation. Two inches long, with a nine-inch tail, Tiny Mite was fed with a goose quill, and he charmed visitors with his silvery fur, his inquisitive face, and his full set of teeth.

Marmosets were deemed so adorable at the time that in Spain women wore them as corsages, and Bostock did the next best thing: He carried the orphan monkey in his pocket. There, Tiny Mite twittered and chirped like a canary. Bostock was pleased with the extremes of his show—Tiny Mite, the two-ounce speck of cleverness, and Jumbo II, the nine-ton war veteran. But being pleased wasn't quite enough; Bostock wanted to experiment. Legend had it that elephants were terrified of tiny creatures, like mice. Would Jumbo really be frightened of Tiny Mite? What if the Animal King could capture a picture of him reacting to the monkey? It would be a great publicity shot.

On Sunday afternoon, August 11, Bostock summoned a photographer to Jumbo's stall. Faced with a space almost too dark to operate in, the cameraman set up a flashbulb and powder and watched as the baby marmoset was brought to the elephant's quarters and put on top of a wooden rail. The cameraman, along with Bostock, waited. And waited. Jumbo didn't see the monkey. Finally, Tiny Mite made a birdlike squeak and Jumbo caught sight of him. Maybe a flashbulb exploded, too. Terrified or panicked, the elephant tried to pull away. His feet, though, were in shackles, so, unable to do more than thrash, he lifted his trunk and roared. Attendants came running and tried to undo Jumbo's chains before he destroyed the stall. Photographs were now out of the question. Worse, in the melee—whether from noise, fright, or hurt—the little monkey died.

It was soon announced that Bostock's show would find another marmoset. This one, meanwhile, would be stuffed.[9]

Tiny Mite was not the only animal Bostock used for an experiment that summer. On another, quieter occasion, and, as Bostock claimed, "in the interest of science," he put his aged Egyptian crocodile, Ptolemy, to a test. At three o'clock in the morning, early in July, two mules pulled a wagon carrying Ptolemy, in a box, to the edge of the Niagara River. Men put the box into a boat, rowed the boat out to an island, and released Ptolemy into the water, secured on top of a log. Bostock and other attendants bet each other on whether or not Ptolemy would survive a trip over Niagara Falls.

As the crocodile approached the rapids near the precipice, the log rolled and threw the crocodile off. Ptolemy drifted on alone. When he got to the edge of the falls, at around three thirty in the morning, observers saw their first reaction from the animal. "With a mighty effort he raised himself partly out of the water [and] waved his arms wildly in the air," one witness noted. "He opened his mighty jaws and closed them with a snap of despair."

Bostock and his entourage moved to the shore below the falls. They scanned the water for over an hour, at which time they assumed that Ptolemy "had been crushed against the rocks and pounded to a jelly." But, just at daybreak, the crocodile's trainer spotted something in the water. He whistled. An exhausted Ptolemy turned his head. The crocodile was "dragged" ashore, where he was tied into a box and hoisted up the high cliff. He was too tired to eat. How long Ptolemy lived after his ordeal is unknown. He was never mentioned again.[10]

IV
SMALL FRY

As days at the fair rolled into the fall, and the humidity yielded, bit by bit, to drier air, excitement at the Exposition became palpable. Ticket sales were up, trains were booked, and boardinghouses were beginning to turn people away. "The floodgate has set in," announced a reporter. He didn't mention Chicago—that dream seemed to be slipping by. Still, officials hoped big crowds would let them hold their heads high at the end of the season.

Less than ten days before the president was to arrive, though, there was an unexpected hitch. Fairgoers on the North Midway began to complain that the Lane of Laughter had started to smell. Visitors expected odors of tobacco and dung, but this was more like rotten eggs that had festered in the sun, and it was so heavy and strong that some guests began to feel sick. Concession owners and fair directors fumigated show buildings, cleaned out sewers, and sent a torrent of water down the main Midway street. Health officials arrived and studied the building plans. Nothing. A week went by. Finally, the source of the nauseating odor became obvious. The asphalt—the pride of Buffalo—had erupted in large sores. One of the spots, near the Filipino Village and the Popcorn Concession, had grown to sixty feet by fifteen feet.

Engineers sampled the putrid spots and came up with a chemical answer: A layer of slag, a waste product of steelmaking used as a bed for the asphalt, had been improperly burned. Its carbons and sulfates were eating Midway streets. Matters would get worse before they got better. The slag had to be dug up and replaced, and, for the time being, the North Midway would feature a stinking hole. The president, with his sensitive wife, was due soon.[11]

On August 17, newspapers printed detailed accounts of the president's schedule. The McKinleys would stay at John Milburn's house on Delaware Avenue, and the chief executive would spend September fifth, President's Day, touring the fair, giving a speech, meeting foreign commissioners, and capping the event with the Illumination. The next day, he would travel to Niagara Falls. With two days to go, papers offered even more detail. The information was important, the *Commercial* explained, because "the president's every move will be followed with interest." On Friday, September sixth, after touring the falls, the president would hold a short reception at the Temple of Music on the northwest corner of the Esplanade. Even though McKinley would be surrounded by mounted and walking escorts throughout his entire visit, Exposition promoters urged well-wishers to show him rousing enthusiasm.[12]

Exposition guards and Buffalo police likely looked to the president's visit with confidence. For the previous four months, the Pan-American grounds and the city itself had been relatively free of serious crime. The superintendent of police in Buffalo, William S. Bull, had worked hard to achieve these results. In April, he had delivered a public warning about the impending intrusion of thieves, housebreakers, shoplifters, pickpockets, and tricksters. City residents, he had explained, needed to be on alert. If a parade went by, they might be drawn to their front porches to see the goings-on. A thief could move to the back and try to get through doors and open windows. At the fairgrounds, too, criminals might be very sly. Male pickpockets might be "well-groomed and gentlemanly," and crooked women might be "humble in attire, unassuming." Don't be too curious, he warned. Don't make friends with strangers, and "Don't be too sure that a crook cannot do you."

Once the fair opened, Buffalo police scoured the environs for suspicious types. They apprehended Sarah Gawley and Minnie May, pickpockets, and Robert Tasnow, alias Dutch Bob, a con man. They arrested Dick "High Card" Taylor, and Hoppy Loftus, a stickup man. And they nabbed Thomas "Spotty Wing" Cullors, who had come from Cincinnati to open safes. Mary Smith, known around town as "the woman in black," who specialized in robbing apartment houses, was arrested as well. There were dozens of others, but, all in all, there were no "big jobs," and most of the confidence men were "small fry."[13]

V

NIEMAN

On the last day of August, a young man called on John Nowak, the proprietor of a Buffalo boardinghouse, and asked about renting a room. He offered a reference—a Mr. Dalkowski, who had just left the day before. He had actually never met Dalkowski, but Nowak didn't know that, and he didn't bother to check. The man seemed harmless. So neatly dressed and groomed, too. Nowak and his wife wondered whether he was a waiter.

Some lodgers later recalled that the young man, who was Nieman, made some friends at the boardinghouse, particularly a socialist named Stutz. The two men had talked for hours. But Nieman didn't speak much to Nowak. "You might ask him a question," remembered the landlord, "getting a few short words for answer sometimes and then again he would ignore you." Actually, Nieman didn't think much of Nowak. Later on, he referred to him as "an old pumpkin-head."[14]

It is hard to know exactly when Nieman began planning to kill the president. He told the police it was sort of an impulsive thing, but he was often confused about dates—and he was awfully good at lying. Two men who had crossed paths with him earlier in the summer claimed that he seemed to have vague plans even then. His older brother, Waldeck, with whom he was as close as anyone, recalled that he began to sound increasingly desperate, declaring, "I can't stand it any longer." Another Cleveland man remembered Nieman saying something similar: "Something must be done."

Whether Nieman's plan had been gestating for weeks or months, or had been born impulsively in the first days of September, the roots of his disturbance lay deep. Since the mid 1890s, when he was in his early twenties, he had been sick. Complaining of coughs and congestion, he visited as many doctors and took as many medicines as his savings would allow. He used drops and teas and atomizers and smoked the leaves of a plant. He left his work at a wire mill because of illness and returned to the family farm.

Among his many troubles was certainly some sort of mental illness. Not only did he worry obsessively about his ailments; he also washed and groomed himself compulsively. And he wasn't trying to please anybody, either, for he avoided people. According to a sister-in-law who lived with him, he "acted very queerly." Nieman was also depressed. He said he was "tired of life" and in July, just before he left home for good, he spoke of an ending. He had demanded money— his share of the family farm. He needed it, he said, right away. "What can you want the money for?" Waldeck had asked. "Look," Nieman had replied, gesturing to a tree that was losing its leaves. "It is just the same as a tree that commences dying—you can see it [isn't] going to live long."[15]

Nieman's illnesses, whatever they were, fueled his distress over people without power—working people, mostly. This was no abstract concern. After he stopped attending school at fourteen, Nieman labored steadily for ten years—first in tobacco stores in Michigan, and then, when his family moved to Pennsylvania, at a glass factory,

Leon Czolgosz, alias Fred Nieman.

processing molten-hot bottles. When his family transferred to Ohio, he took a job at a wire mill in Cleveland and worked ten-to-twelve-hour shifts. Although rarely cited for carelessness, he had mishandled a sharp wire on one occasion and bore signs of the slash on his cheek. Then, in 1894, he was done. When he and Waldeck joined others in a strike for better pay, the men were fired.

By all accounts, the loss of his job was something of a personal watershed, the moment when Nieman exchanged his Catholic faith for political purpose. He and Waldeck had been urged by their priest to pray for help during this troubled time, but, his brother explained later, "no help came." As the two men lost confidence in the Catholic Church, they found a degree of comfort in the socialist labor party in Cleveland. Nieman began to buy books and pamphlets about Christian "superstition," including the atheist magazine *The Freethinker*. He also liked Edward Bellamy's bestselling *Looking Backward*, a romantic utopian novel that described a world without strikes, where property was shared.

Although Nieman was eventually rehired at work, he couldn't stick it out. In 1897, after two years on the job, he suddenly quit, saying he was sick. His stepmother called him a malingerer. Waldeck took him at his word and suggested he go to a hospital. Nieman had an answer for that. "There is no place in the hospital for poor people," he argued. "If you have lots of money you will get well taken care of!"

So he stayed at home on the farm and "fussed about," fixing machines, clocks, and wagons, and shooting rabbits and squirrels. He was a very good shot, noted Waldeck, with a revolver.

In the early summer of 1900, Nieman came across the newspaper article about Gaetano Bresci's assassination of Umberto I. Less than a year later, he discovered the anarchist Emma Goldman, when she spoke in Cleveland. In her lecture, Goldman had talked about how the "galling yoke of government" made it hard for individuals to shape their own futures and their careers. "Anarchism," she explained, "aims at a new and complete freedom." Churches were oppressive, too, she said, and even modern schools inhibited freedom. Her anarchy did not support violence or bomb throwing, but she knew others—including those with "high and noble motives"—who felt differently. Nieman liked what he heard. Listeners recalled that he had applauded vigorously.[16]

Goldman was to Nieman both fire and balm, and, shortly after hearing her, he sought out other anarchists, including a man named Emil Schilling in Cleveland. Nieman talked to Schilling about "forming plots" and asked about the assassin Bresci. Were there, he wondered, similar plans afoot in the United States? Bresci, after all, had been "selected by the comrades to do the deed." Schilling told Nieman that his associates did not do any such plotting. He thought Nieman seemed odd, and suspicious. But it was only later, Schilling said, that he realized his visitor had something very particular in mind.

Nieman tried to see Emma Goldman again, traveling to Chicago when he left Cleveland in July. He followed Goldman to talk with her, but she brushed him off—she was leaving town to visit the Pan-American Exposition. She did, however, introduce him to a friend of hers, an anarchist editor named Abraham Isaak. Isaak recalled that Nieman seemed especially angry about "the outrages committed by the American government in the Philippine Islands." Nieman felt they did "not harmonize with the teachings in our public schools about our flag." The young man had a request, too. He was, said Isaak, intent on "soliciting aid for acts of contemplated violence."

Isaak said he offered no support. In fact, he claimed he was put off. Nieman's questions and ignorance suggested that he was an impostor or a federal agent. The editor published a notice against him on September first, describing how a well-dressed, narrow-shouldered man was making the rounds in Cleveland and Chicago. "Comrades are warned in advance," it read. Isaak was either covering his collusion with Nieman or, more likely, he was nervous about the young laborer. Either way, the warning was delivered too late. By September first, Nieman was back in Buffalo.[17]

VI
CUBA LIBRE

Fred Nieman was the most dangerous man circulating through the Pan-American grounds in late summer, and he stood on the end of the spectrum of sanity. But he was hardly the only McKinley critic at the fair. The Exposition swelled with others who, quietly or unremarkably, disagreed with the president. Even John Milburn had at one time voiced his concern over McKinley's imperialist policies.

Then there were dissenters who were more vocal. On Cuba Day, August 29, just a week before the president was due to visit, Cuban dignitaries gave Exposition audiences an earful. It wouldn't be the first opportunity for fairgoers to hear from Latin American visitors, of course. By showcasing New World republics, the Exposition had offered its exhibitors a stage from which to talk to the United States all season long, and they had done so—at building dedication ceremonies, banquets, and other formal events. Many of them spoke for the business communities in their countries—those invested in modernizing, in selling products, in attracting immigrants and capital. Some of them glorified their indigenous pasts, and others emphasized their affinities with Europe. They did not speak with one voice.

All of them, though, were certainly aware of how the Pan-American represented a departure from earlier world's fairs. Latin American countries had been allocated limited space at previous Western expositions, and they frequently had been grouped together in one pavilion and located near colonial displays. Catalogues and official reports had sometimes barely acknowledged their presence.

At Chicago in 1893, individual New World nations had been better recognized, but most had been placed in generic buildings such as Agriculture or Manufactures. Furthermore, they were asked to demonstrate the "aboriginal" end of the progression from savagery to civilization in anthropology exhibits. Peru and Bolivia, for example, had been urged to round up "wild" Andeans for display. When some of these indigenous people arrived in the Windy City wearing Western clothes and haircuts, they had been pressured to look "primitive."[18]

Buffalo, like Chicago, emphasized the advanced status of the United States, but it also encouraged Latin American republics to build exhibit halls (next to state exhibits) and to show off accomplishments in the arts, literature, education, and commerce. Sometimes this worked out, and sometimes it did not. In July, the Exposition suffered a public-relations setback when the keynote speaker for Chile Day, Don Carlos Morla Vicuña, minister to the United States, caught pneumonia. He languished for almost five weeks in a Buffalo hotel, rallying, then sinking again, and died on August 20. For Mexico, it was a matter of dignity more than grief. Mexican president Porfirio

Diaz had urged that the Streets of Mexico Midway exhibit "not in any way bring ridicule upon Mexico, her inhabitants, or buildings." Diaz's protests went unheard, though, and the "Streets," with dancers, burros, and peasants, set up shop. The Mexican president could take pride, though, in Mexico's handsome two-story building, its celebrated military band from Mexico City, and Captain Garcia Cuellar's visiting soldiers.

The Pan-American Exposition not only provided New World republics with exhibition space; the local and national press printed geographies and histories of visiting countries and transcripts of speeches by Latin American businessmen, politicians, and leading educators. Director-General Buchanan, fluent in Spanish, translated a number of speeches himself. Depending upon what newspaper they read, readers might learn for the first time about Latin American history and cultures.[19]

But it was the dedication ceremonies on special "days" that attracted the most attention. And Cuba Day offered visiting islanders a chance to celebrate their nation's past and future and to give their hosts a lecture about current events. It was the first time Cuban officials had spoken formally in the United States since the War of 1898.

Late in the morning, in the cool spaces of the Temple of Music, the Buffalo audience mingled with what one reporter noted as "dark-skinned, dark-eyed foreigners." Honorable Daniel Lockwood, representing the Exposition, opened the ceremony, praising Cubans and announcing that "every American felt proud of the fact that the Stars and Stripes have stretched their protecting folds over the island's people." Cuban speakers offered some polite deference in return, noting the "greatness and grandeur" of the United States and declaring that there was "not a man in Cuba who does not feel a profound sense of gratitude for the United States."

At some point, though, the light-skinned people must have straightened in their seats. Edelberto Farrés, president of the Cuban

Commission to the Exposition, wanted to correct some misperceptions. The recent war against Spain? The war that, to many Americans, seemed a quick rout? In Cuban minds, he said, it was a long war, fought for many years on Cuban soil, by Cubans themselves. The fact that Cubans had managed to put up a building and put on a big display at Buffalo revealed their resilience. "It is surprising," Farrés said, that, "after a war of devastation which cost the lives of nearly 50 per cent of our population, we were able to make the showing we have at your grand exposition."

Farrés also addressed political matters. He noted that the ink was barely dry on the Platt Amendment. This legislation, signed with many misgivings by the Cuban government, had given the United States the right to interfere in Cuban affairs. Cubans, the United States Congress had argued, needed American guidance and protection. Farrés demurred. We have, he declared, "since this exposition opened proved that we have the courage and the progress and the ability to govern ourselves."

A Cuban school commissioner named F. R. Machado had more to say. People were under the impression, he said, that Cubans were never happy, and that they rejected the involvement of the United States. The matter, he said, was not hard to explain: "Chains, gentlemen, even golden, are always chains, and it is nothing particular that those who have been deceived and tyrannized for centuries should be suspicious. . . ." Then he used another analogy. "We are, I think, very much like champagne." Champagne, if "too closely confined . . . will always strive to break the bottle. If it does break it by much pressure, then those around it may be hurt."

The Cubans were not done. Dr. Louis A. Baralt, a Cuban orator and linguist, said he was glad that the United States and Cuba were getting acquainted. "But," he added, "all must remember that this country [the United States] is not America." He also suggested that there was an implicit bargain at work in the Exposition. "We shall sit

at the same banquet table and shall be one people, one great brother-
hood in common interest . . . [but] we have taken [you at your] word
in saying that Cuba ought to be, and shall be, free." At least one section of the audience cheered heartily. One observer said that this event
perhaps had "more significance than any other which has been held
at the exposition."[20]

Just over nine days later, President William McKinley, to whom
many of these remarks were addressed, would stand in the very same
cool and airy rotunda and stretch out his hand to greet well-wishers.

4.

The Blood-Colored Temple

I
GOODBYE TO OHIO

The day before they left home for the Exposition, President and Mrs. McKinley kept their calendar open to pack and organize, and took no callers. The packing would not be arduous—it was a relatively short trip of ten days. First the Exposition, then a visit with good friends in Ohio, then a Grand Army of the Republic veterans' encampment in Cleveland. William McKinley liked these veterans' reunions—a chance to reconnect with comrades from those hard days long ago.

They made one exception in their empty schedule. They drove to the Stark County Fair for Children's Day, toured the grounds in a carriage, and stopped to greet groups of children. Many of the younger folks, said one reporter, "had never before seen a real live president." The McKinleys liked children; they had never seen their own grow up. They had lost two daughters, three-year-old Katie from typhoid in 1875 and Ida, named after her mother, just four months old, in 1873. Every Sunday when they were in Canton, they took fresh flowers to Westlawn Cemetery to place on their graves.

The McKinleys left Canton for the Pan-American Exposition at ten in the morning on Wednesday, September fourth. They took with them members of the Secret Service, a nurse, a maid, three nieces, two stenographers, and George Cortelyou. In addition, the White House physician, Dr. Presley Rixey, traveled with the party. Dr. Rixey, a former naval surgeon, had been with the McKinleys since 1898, and he now hovered over Ida. Although she had regained her health during the quiet days in Canton, he wanted to be at the ready should she suffer one of her frequent "turns."

The McKinleys' train traveled through Cleveland, where people lined the tracks to see the president. He obliged, and, standing on the platform at the rear of the train, he waved good-bye to Ohio. The train passed Erie, Pennsylvania, at 3:45 p.m., and, an hour later, at Dunkirk, New York, a greeting party from the Exposition climbed aboard.[1]

People who believed in omens pointed to the president's arrival in the city of Buffalo as a portent—as though the strange thing that occurred should have been enough to turn the train around and send it back to safety. The presidential entourage, in three railroad cars, arrived at Buffalo's Central Station just before 6 p.m. The city had planned a powerful greeting for the dignitaries: simultaneous blasts from ship horns, factory whistles, church bells, and cannons, along with hurrahs from well-wishers on the streets.

The train arrived, belching smoke and steam, and the noisemakers went to work. How much of the clamor the arriving guests could hear is unknown, but soon after pulling away toward the Exposition, they felt a sudden blast, this one close and loud. An explosion shook the train, shattering the Pullman windows to "smithereens" and blasting out windows on nearby buildings.

We can only imagine the anxiety of Presley Rixey, thinking of how little it took to put Ida McKinley over the edge, or the fears of George Cortelyou, readily picturing the president's enemies. Another witness shared Cortelyou's concern, for, seconds after the blast, he shouted something about anarchists. Others, too, panicked. One had seen a "swarthy" man in the vicinity.

There were no anarchists, though, this time. And there were no dark-skinned suspects. The culprits, soldiers of the Seventy-third Coast Artillery, had misfired a twenty-one-gun salute. The cadets had been ordered to fire vintage guns taken from a War Department exhibit in the Exposition, even though most of them had never seen, much less operated, muzzle-loading guns. Nor had they been instructed in the positioning of the guns or the proper timing. Not surprisingly, the timing was off, and they had fired close to the presidential train.

With nothing to repair beyond military pride, the McKinley train steamed the rest of the way to the Exposition. The presidential party, now dressed in evening black, transferred to a carriage and made its first tour of the fairgrounds. They then rode down Delaware Avenue, passed the big houses draped in patriotic bunting, and settled in at John Milburn's house for the night.[2]

Between the first day of September and the fourth, Fred Nieman entered and exited Nowak's boardinghouse, not making any fuss about where he went, what he saw, or whom he met.

He went down Broadway to a barbershop and asked about a bottle of glycerin. Maybe he wanted to tame his hair, or he went in for a shave and a haircut. It would be just like Nieman to be fussing with his looks. Maybe, though, it was something else altogether. Take glycerin, add a little nitric acid, and the result wasn't a beauty product. If Nieman had bomb-making in mind, he hadn't studied his ingredients. The bottle that he stole from the barbershop contained nothing but cod liver oil.

He was more careful about another item. In Walbridge's Main Street store, he found the perfect firearm: a .32-caliber Iver Johnson revolver. Its barrel was short, which was handy. But Nieman didn't worry too much about the specifications. He was already convinced it was the gun to do the job. Gaetano Bresci had used this very model in Italy.[3]

Nieman carried resentment with him like a malignancy—it ate away at him, and he could not escape it. Being at the Pan-American fair, walking to and from jobs, probably fueled his feelings of injustice. In fact, for a man who believed that states were tyrannical and that all rulers, including American presidents, were oppressive, there was probably no place in the world more likely to stir his fury.

Imagine Nieman walking into the Exposition grounds over the Triumphal Bridge. This strutting edifice, hung between four massive pylons, served as a colossal tribute to state power. More than a hundred feet off the ground, men on rearing horses rose over trophies of war and claimed victory for American republics. Copper shields, coats of arms, and banners, all suspended on cables, honored the nations of the Exposition and their military might.

Imagine Nieman wandering across the Esplanade into the United States Government Building, its blue dome crowned by the symbol of victory. Inside, set atop a gigantic model of the globe, Uncle Sam's boats sailed over the world. More than three hundred little lead ships, including gunboats, submarines, torpedo boats, and battleships, occupied almost every ocean. Attendants moved the miniature fleet every day, so visitors could stay current with the nation's imperial reach.

Another pavilion was dedicated to that reach. A three-thousand-square-foot exhibit was devoted to displays of curios and implements of the Philippine islands. It included farming tools, fishing nets, clothing, and household items. And it featured weapons that United States soldiers had taken from Filipino fighters after they were killed or captured.

Did Nieman see the Midway? He didn't seem the type. If he had,

The Triumphal Bridge on Flag Day. The 100-foot-long American flag,
hoisted between the pylons, was said to be the largest in the world.

he might have struck up conversations about politics and power with
people who performed in the living exhibits. While there is not a
shred of evidence that he met Pablo Arcusa, he likely would have
agreed with his desire to rid his native country of Americans.

Nieman professed to be as concerned about poor people as he was
about the heavy hand of government. Where some at the turn of the
century took pride in American successes and victories, he saw only
struggle. And, in his distress, he found somebody to blame. "McKinley was going around the country shouting prosperity," he said, "when
there was no prosperity for the poor man."[4]

What did the fair say to Nieman about money in America? The
Exposition showed off imperial power like a jeweled ball gown, but it

did not parade or flaunt individual wealth. That would have been un–
American. In fact, the Republican press claimed that Rainbow City,
like other fairs of the time, was friendly to working people. It was
"the people's fair." Newspapers reported that in the winter of 1899,
when the city began asking Buffalonians to invest in the fair, more
than eleven thousand residents had pledged $1.5 million. Subscribers
to the Exposition, who had bought shares for as little as ten dollars
each, represented a wide swath of the city.

The Pan-American Exposition also offered jobs to thousands of
local people. There were problems, of course. Companies found ways
around hiring local men, and the site saw its share of walkouts. Plas–
terers and ironworkers and carpenters all went on strike for better pay.
The good news for workingmen, though, was that labor was well
organized in Buffalo.

On Labor Day, September second, in fact, a huge parade of fif–
teen thousand workers testified to the importance of Buffalo labor
unions. Papers pronounced the scene a reflection of the new era of
prosperity, where everybody "marched to the music of good times."
Many of the marchers also cheered on Samuel Gompers, head of the
American Federation of Labor, when he delivered a speech in the
Temple of Music. Gompers thanked the Exposition leaders for their
reception and praised the men who built and operated the fair. He
reminded his listeners that unions furthered justice and prevented
strikes. Exposition officials, such as President Milburn and Director-
General Buchanan, thanked him, politely.[5]

Local working people not only put up the fair and kept it run–
ning; they also visited the display buildings, walked through the
Midway, whooped on rides, and took pride in the power of the
nation along with thousands of other guests. But doing so cost
money and took work time. Did the scrubwomen, cooks, carpen–
ters, and custodians have the means to take in many of the sights?
They saw the grounds, to be sure, but it was at the edges of the day

when they swept the asphalt clean, scoured floors, repaired exhibits, and strained the fountains of litter. How much were they able to tour and revel in Rainbow City?[6]

The year before the fair opened, a wealthy Buffalonian suggested that laborers save money so that when the fair rolled around, they would have enough cash on hand to attend. Businesses, he said, should dock paychecks twenty-five cents each week. Saving money would bring about a double benefit: fun at the fair *and* disciplined habits.

That idea never caught on, and it is unclear how hard fair directors worked to open the fair to the very poor. They did cut ticket prices on Sunday afternoons, from fifty to twenty-five cents. But visitors couldn't see the Midway then, because its content had been deemed inappropriate for the Lord's Day. And even on Sunday it was hard to see the fair the "right" way. The grand entrance to the Exposition, from the south, was designed for horses and carriages, not trolley riders. Good seats for concerts and fireworks also cost extra, and even restrooms charged for towels and soap. Women without means also likely never saw the interior of the Women's Building. It was not meant for public resting or lounging.

Rainbow City seemed to open its gates widest, then, to men and women who had money to spend and days off to enjoy it. On Labor Day, when workers had a rare weekday vacation, many families saw the fair for the very first time. These women and children hadn't been disinterested before, certainly. Some of their fathers and husbands had even helped build the Exposition, and they toured the grounds listening with pride about how exhibits had been put together. What they had been missing for the previous four months was not the motive to see the beauty and wonders of Rainbow City, but the means.[7]

II
POMP AND CIRCUMSTANCE

On the morning of President's Day, Thursday, September 5, gondoliers staged a last-minute walkout, leaving the fair's picturesque waterways empty. Pan-American enthusiasts ignored the strike, however, and threw themselves into the red, white, and blue blur of presidential pomp. They enjoyed a day free from cannon accidents, too. The Seventy-third Coast Artillery, scheduled to fire off big guns again, had been positioned as far away from the Lincoln Parkway gate as possible, "so that the salute will not disturb the presidential party."

As for President McKinley's trusting himself to the public? Guards seemed to be everywhere: Secret Service men stayed close by, soldiers surveyed the scene from horseback, and police detectives and Exposition police circulated through the crowds. Other dignitaries—senators, foreign officials, and ambassadors—mingled together in what one visitor called "a grand array of might and power."

Arriving at 10:30 a.m. on President's Day, Mabel Barnes and Abby Hale struggled through the masses to see the chief executive. They were pleased to have "first class views" of McKinley as he reviewed the troops at the stadium. They skipped his speech, instead strolling through a forested island in one of the Mirror Lakes and making a cursory visit to the Ohio State Exhibit. They then made a beeline for the part of the Exposition they loved the most: the Midway. It was Mabel's twenty-second visit to the fair.[8]

Late in the morning, McKinley gave an address to fifty thousand people from a covered Esplanade bandstand. The motion-picture crews filming the event did not have the technology to capture the vibrant colors: the purple swags and drapes on the stage, the scarlet-coated musicians, and the blue robes of Chinese diplomats. They got

14563   Crowds at the Temple of Music where President McKinley was assassinated, Pan American Exposition.

Stereographic view of crowds on President's Day.
The Temple of Music is in the background.

a good look at the crowd, though, with its straw hats and parasols stirring like whitecaps on a human sea. And they filmed the president, taking in his big bowtie and his wide white shirt and capturing him bobbing with emphasis.

"Expositions are the timekeepers of progress," McKinley began. "They record the world's advancement." They stimulated enterprise, he said, and produced friendly rivalries and, in this case, demonstrated the development of the Western Hemisphere. Soon enough, the president steered into his favorite topic: trade. With the world smaller than ever, thanks to fast trains and ships, there was less likelihood of distrust and more room for trade. United States workers, producing more than ever in this climate of "unexampled prosperity" just needed more markets. To help them, the country needed more ships, an Isthmian canal to the Pacific, and a cable to new outlets in Hawaii and the Philippines. McKinley saluted the father of inter-American commerce, former Secretary of State James G. Blaine.

Below him on the fairgrounds, the sun beat down on the audience, and here and there women fainted and dropped. Ambulances darted across the asphalt. Closer by, the crowd clapped at intervals. Some of them, especially those at the very front, could actually hear what the president said.

McKinley closed his speech by appealing to the future. He hoped Rainbow City would not only stimulate commerce and international respect but also serve as a beacon. "This creation of art and beauty and industry will perish from sight," he said, "but [its] influence will remain." The president sat down, held the hand of his wife, and then rose to acknowledge the cheering crowd.

The speech was swiftly carried by telegraph. Republican editors applauded the talk of opening Latin American markets, while in Europe, the response was muted if not hostile. McKinley's "Pan-Americanism" fell flat in London. It was "imbued with an expansive, even aggressive spirit," commented London's *Standard*. The imperial United States was determined to go its own way regardless of the consequences. "Europe will never march out of America," the paper declared. Other foreign newspapers echoed the sentiment.

Not far from the podium, with his gun heavy in his pocket, Nieman watched the president speak. He had been at the fair for hours, seeing visitors massing for the distinguished guest. "I heard it was President's Day," he said later. "All those people seemed bowing to the great ruler." He was now set on killing McKinley. "It was in my heart; there was no escape for me," he confessed. "I could not have conquered it had my life been at stake." But he did not think the time was right.

Police detectives, who dutifully surveyed the throng, did not take notice of Nieman. Instead, they spotted an older woman trying to climb up on the president's stand as he finished his speech. They blocked her, and she swore loudly. Some people mistook her for Carrie Nation, the outspoken temperance advocate. Later in the day, officers noticed another dubious character—a young boy who had been

President McKinley addresses the crowd
at the Exposition, September 5.

trailing the president and who carried a bundle under his arm. Obviously, said a reporter, he was "a crank or a dangerous person." Having followed McKinley from building to building, the boy was finally stopped. The bundle that he was carrying? His dinner.[9]

Throughout the grounds that afternoon, the crowds surged. Entrance turnstiles that had revolved slowly in the morning began to spin, and by the end of the day recorded 116,000 visitors. The final tally must have warmed the hearts, not to mention eased the minds, of Exposition backers. A week earlier, directors had predicted that if daily numbers persisted, the fair would be able to honor its debts and reward investors.

Over on the Midway on President's Day, Frank Bostock was in his element. He had promised audiences that he would make a rare appearance in the training arena, and there he stood, with Wallace, Jr., one of the "intractable" lions. Bostock took a long whip, and while Wallace skulked from one end of the circular cage to the other, Bostock threatened him. He did not have the lion do tricks. "No one can ever do that," said his press agent, but the Animal King did make the animal back down in fright. His "calm and complete victory," reported the agent, "earned him a storm of applause."

Chiquita, on the other hand, was lying low. Even on Cuba Day, August 29, she had gone about her business without comment. Cuban dignitaries had come, toured the grounds, and delivered formal addresses, but Chiquita, the most famous "Cuban" at the Exposition, was not seen or mentioned. Perhaps she did not want to call attention to her fictitious descent. Or perhaps she wanted to avoid attention altogether. She and Tony were still seeing each other. He had been hired by the Indian Congress on the Midway after Bostock fired him, and had secured the help of a showman's child to deliver notes to her. Some nights he even managed to climb through the back window of her quarters to spend time with her.

William McKinley spent the rest of Thursday, September 5, touring foreign exhibits, accepting cigars, and talking with dignitaries. At the New York State Building, he enjoyed, or appeared to enjoy, a luncheon of crabmeat pâté, sweetbreads, roast turkey, and cream pudding with cherries and chestnuts. Fortified by a short nap and bolstered by her nurses, Mrs. McKinley had rallied enough to accompany her husband as he spoke from the bandstand. But then she wilted. The Board of Women Managers had prepared a lobster Newburg luncheon for her, but she was "so weary" after her husband's address that she retreated to the Milburn home to rest. Amid table settings of white roses and asters, the women lunched on lobster and bonbons without her. The guest list was limited, claimed one observer, and, even without the guest of honor, the occasion was deemed quite satisfactory.

In the evening, for the benefit of the distinguished visitors, fireworks master James Pain lit the sky with the best displays gunpowder could produce. For two hours, he fired off his magical mix of colors: mauve and magenta, canary yellow, Nile green, vermillion, and azure blue. He shot up violets and pansies, passionflowers, heliotrope, and clusters of fireflies. He let loose Whirlwinds, Sunbursts, Peacock Plumes, and even Niagara Falls, with a cascade of liquid fire. To honor the president, Pain lit rockets that screamed like an eagle; then he produced in lights a naval battleship and McKinley's own magisterial face. His grand finale was a shell celebrating "Our Empire." At 1,000 feet in the night sky, a firework burst open with the colors of the United States, Cuba, Puerto Rico, and, of course, the Philippines.

Mabel Barnes and Abby Hale made sure to get a good seat at the evening's display, so they arrived at 5:30, two and a half hours early. They had a picnic, rested, and studied the crowds. Being such early birds, they also got a good look at McKinley as he gathered with his entourage at the Life Saving Station on the Park Lake. They could not hear the conversation near the president's group, however.

Which may have been a good thing. McKinley's companions were uneasy. Between the bursts of colored light, the night seemed pitch black. "I was impressed," said Secretary of Agriculture James Wilson, who was sitting near the president, "with the ease with which some evil disposed person could have crept up in the darkness . . . and have done the President bodily harm." George Cortelyou felt the same way. As the two advisers sat on the benches watching the fireworks, they discussed how they might protect McKinley. Wilson found the topic so consuming that he could take little delight in the spectacle. The reason, he said later, was "dread."

Perhaps the president's escorts wanted to let the world know that they had been alert that night. Or perhaps, out of the corner of an eye, they had actually seen someone—the same slight, brown-haired man—everywhere they went. And then they had put it out of mind. Nieman confessed he had been following McKinley but couldn't get a good look at him. So he waited. He wanted "a better chance."[10]

III
THE WAITER

Some of the people who were in the Temple of Music at the Pan-American Exposition on September 6 fell into shock for a month or so, and then they picked up their old routines. Others, though, saw their customary habits shattered forever. For these people, there was always life before Friday, September 6, and life after.

Jim Parker knew about that split. He was among the thousands who hoped to get a look at the president, or, even better, to shake his hand. Parker was a waiter at the Plaza Restaurant, just north of the Electric Tower. Before that, he had been a waiter at Saratoga, New York, and even earlier he had worked in Atlanta and Savannah, in the post office, on the railroads, and as a constable. While it is likely none of these jobs paid particularly well, all of them were preferable to the work he had done as a child. Until he was eight years old, he had been enslaved in Georgia.

Parker may have actually felt lucky to get a job at an Exposition restaurant. While service jobs had been an African American domain for decades, young white women recently had been offered more and more of these positions. But Parker had experience, which probably kept him employed. And he was hard to forget. His slender face and friendly eyes were not particularly memorable, but in an era when many men barely topped five feet tall, his height—six feet two inches—was worthy of notice.

Nieman might have said he cared about people like Parker— those who had almost nothing and who faced a future with only a little bit more than nothing. These were the men, on the bottom rung of America's ladder of power, for whom anarchists spoke and for whom violent anarchists sup- posedly did their killing. Parker, though, would not have returned the sentiment. He may not have liked how hard he had to work, but whatever anger he stored was not directed at the American pres- ident. Other African Americans were disappointed with McKinley.

Jim Parker.

Not only had he seemed to turn a blind eye to the atrocities of lynch- ing, but lately he seemed to be getting cozier with ex-Confederates.

In December 1898, in Atlanta, McKinley had affixed a gray badge to his lapel as a sign of forgiveness to his former enemies and announced new federal oversight for Confederate, as well as Union, graves. There had been, McKinley announced, "an evolution of sentiment." While many black leaders in that brutal era recoiled at McKinley's will- ingness to forget and forgive, more conservative spokesmen, such as Booker T. Washington, continued to stand behind him.

Whatever the nature of his own politics, Jim Parker felt enough respect for the chief executive to want to meet him. In the middle of the afternoon that Friday, he slipped away from the restaurant, snaked his way through the Esplanade crowds, and got in line near the open door of the Temple of Music.[11]

IV

NIAGARA

Ida McKinley had been to Niagara Falls before, but her husband had not, and the president beamed with pleasure on Friday morning as she pointed out the celebrated sights of the cataract and the gorge. The presidential carriage drove through the nearby village, and several young boys raced alongside, took off their hats, and gave them three cheers. Other onlookers took up the cheers, and soon the whole crowd shouted its enthusiasm. The boys were thrilled to be so close to the McKinleys, and the couple, in turn, was enchanted.[12]

There were other people in Niagara Falls that day who wanted President McKinley's attention. It is unclear how or why they came up with the plan, but two vaudeville actresses, Martha Wagenfuhrer and Maud Willard, decided to capitalize on the president's visit. The women, who performed in local theaters, decided that they would climb into barrels (outfitted barrels were, it seems, a local specialty), enter the Whirlpool Rapids below the falls, rotate through the Whirlpool, and dazzle the president. If they lived to tell the tale—and no woman had ever done so—their celebrity would be assured and they would have a new show to sell.

Thirty-four-year-old Wagenfuhrer, a recent immigrant from Germany and a resident of Buffalo, would go first. If the timing was right, her barrel would catch the eye of McKinley as he looked down the gorge. If he missed her, she would at least have his crowds, and her

 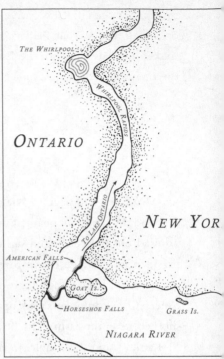

Martha Wagenfuhrer (*left*); a section of the Niagara Escarpment, from
above the big falls to beyond the Whirlpool (*right*). The river flows north
from Lake Erie, over the two falls, and toward Lake Ontario.

stunt would be witnessed by thousands of people clustered on the
Canadian and American shores.

People would later say that Martha had been drinking, and that was
the reason she delayed her start. But there were other problems. The
men carrying her barrel down the steep path to the river's edge lost
their grip, dropping it. It rolled down the bank, slamming into trees
and rocks. When Wagenfuhrer saw that its hoops had come loose, she
insisted they be fixed. A cooper was summoned. Hours passed before
she slid into the big cask, and by then President McKinley had been
to the Whirlpool, seen the gorge, marveled at the falls, had lunch, and
toured a power plant. He had even made it back to the Exposition. To

make matters worse, the people lining the railroad tracks above the gorge—some of whom had helped pay for Wagenfuhrer's stunt—had turned mean. "Fake!" they yelled down at her. "Coward!"

Men closed the lid of her barrel at around six o'clock, but not before she heard something about President McKinley. No time for that, though. Soon enough, she was bounding through the lower Niagara River. Her barrel drifted toward the rapids and took the current. Up above, in electric cars, people strained to see. The barrel pitched and spun and was swept into the Whirlpool. And there it stayed, going around and around in a sickening circuit, for over an hour.

As the light faded, watchers had a hard time keeping track of Martha, her dark barrel lost against the dark water. Along the Canadian side of the river, rescuers lit fires near the places where the Whirlpool usually disgorged objects, and they waited. Finally the barrel sprung free, and, with a great shout of triumph, rescuers seized it and wrapped it with a towline. Ashore, they broke into the cover and pulled the woman out. Martha Wagenfuhrer was unconscious, but she was breathing. After ten minutes on the shore, she came to.

Traumatized, Martha discouraged Maud Willard from performing her own barrel stunt the next day. "She wasn't as strong as I was," Martha said. "I tried to persuade her to give it up." Maud, though, wasn't going to give up anything.[13]

As for Martha's new celebrity? Her feat was about to be swallowed by a news story coming out of Buffalo.

V

THE COOL OF THE TEMPLE

At the International Hotel in Niagara Falls, on the American side, where the presidential party ate lunch that Friday, members of the group thought that Director-General Buchanan seemed preoccupied.

During the trip on the Gorge Railway, he had also seemed restless, walking back and forth, back and forth, by the touring car. Now he paced the dining room. One of the party asked his wife what she thought was bothering him. Mrs. Buchanan laughed. "Isn't he foolish," she said. "He told me he felt as if something was going to be wrong before they got through."

Chances are that exposition directors during presidential visits always felt themselves perched on the edge of calamity. Buchanan couldn't have known that Nieman had taken a streetcar to Niagara Falls that morning and had tried to follow the president's group. Or that he had tried to get close to them so that he could, as he put it, "carry out his purpose." Or that he had just taken a streetcar back to Buffalo and headed to the Exposition grounds.[14]

After returning from Niagara Falls, President McKinley left his wife at the Milburn house to recover from the excursion and took a carriage to the Temple of Music with his entourage. He had changed his white shirt since the trip to the falls, and, as his carriage inched its way through the inevitable crowd, he appeared crisp and ready. He walked up the steps of the building jauntily—or as jauntily as a heavy man could move in the heat—looked up at the arches above him, and noted aloud that it was pleasantly cool.

Inside, he positioned himself in the sort of place he must have stood thousands of times as a public man: under flags, drapes, potted trees, and palms. He stood amid a small army: three Secret Service men, a handful of fair dignitaries, Exposition guards, and the loyal George Cortelyou. Not to mention almost sixty soldiers of the Seventy-third Coast Artillery. These troops, likely recovered from the ignominy of blasting the president's train, were pleased to be given this position of honor.

The reception line connected the exits, all the better to streamline the greeting. It would be a short reception. The day had been long, the president probably was tired, and Mr. Cortelyou, as always, was nervous. But McKinley was good at this. In fact, smoothly shaking hands with constituents was considered his specialty. He took an individual's right arm, got enough of a grip on the elbow to move the person along while making a pleasantry, and swiveled to the next in line.[15]

Outside, in the sweltering air, some people had been waiting in line for more than an hour. Nieman had been waiting. Despite the temperature, he looked pulled together. He wore a laundered shirt, a vest, a jacket, and he had combed, parted, and oiled his hair. He kept one hand in his pocket and his fingers closed around his handkerchief. His handkerchief, in turn, was wrapped around his gun.

Finally, the door opened and the visitors were welcomed into the rotunda. Inside, an organist worked his way through music. Was it Bach or Schumann? No one could remember later, although other details were etched diamond-clear.

The line was tight, and the Temple of Music was soon dense with bodies. Officials asked those carrying lunchboxes—who knew what was inside?—to go to the rear. Chances were that these uncertain sorts would never make it to the president. A flutter of white stirred among the crowd—another concern. The closeness had begun to make people hot and uncomfortable, so they took out their handkerchiefs. They unfolded them, mopped the moisture from their faces, and put them away. Later on, as they got close to McKinley, they pulled out the cloths again and wiped their hands.

Jim Parker, the waiter, had arrived early enough to get a good place in line, but even now he knew he would be lucky to get to the president before the reception ended. He would have liked to move ahead of the clean-cut man in front of him, but the man had sharp elbows and kept his place. Parker couldn't see this, but the man had pulled out his handkerchief, holding it close to his chest.

Opposite the president, where he had been positioned, Secret Service man James Ireland studied the reception line and watched an injured man take his turn. Poor fellow, three fingers bound up. Had to use his left hand to say hello.

Now a black-haired man with a mustache was taking a long time with McKinley. He looked Italian. Anarchists, the agent thought, looked Italian.

George Foster, another Secret Service agent, also noticed the dark-skinned man.

Together, they moved him along.

It was four o'clock. President McKinley had been shaking hands for about ten minutes.

A mother and a little girl moved into place. The president bent down to talk with the child.

Then, coming up, a neat, boyish-looking, fair-skinned fellow. His hand was wrapped up, too—another injury. A laborer, no doubt. And close behind him, some sort of colossus—a towering black man.

Edward Rice and George Cortelyou exchanged signals. Rice, head of ceremonies at the Exposition, took his watch out of his hand. At the same time, Cortelyou pulled out his timepiece. Ten minutes, they had agreed, was long enough.[16]

As close as anybody could remember, it was seven minutes past four.

It was Nieman's turn to shake hands.

Instead of his palm, he brought forward his revolver.

And fired.

5.

The Emergency

I
SHOCK

To people within earshot, the two gunshots went off so close together they sounded like one. To others, it wasn't the sound of the revolver that registered—it was the look of the president: His heat-flushed face was now white. He straightened up "convulsively," said one witness—and recoiled. Others said no: He stood like a statue, perfectly still.

They agreed that he did not fall.

And this meant that he could be shot a third time. Nieman, his cotton handkerchief now singed, still held the revolver.

Behind him, Jim Parker's rage rose in an instant. "You son [of a bitch]," he shouted, as he grabbed the assailant around the throat and pushed him toward the floor. Secret Service men jumped into the act, knocking Nieman down and grabbing the gun. The Artillery soldiers rose to the occasion, too, but . . . it just wasn't their week. The hand they grabbed did not belong to Nieman but to one of the agents—probably Albert Gallaher—and, attacking him, they took away the weapon. The floor became a blur of fury—six to ten men "struggling and swaying."

The president himself was calm. Eyeing the prisoner and the melee on the floor, Gallaher heard him ask the agents to "go easy on him." George Cortelyou, meanwhile, had found McKinley a seat, and a

Washington reporter named Randolph Keim stepped in to help. "I began fanning him with my straw hat," Keim remembered. He could see McKinley's opened vest and powder marks around a bullet hole. Blood had begun to seep through. The president noticed it, too, but he remained composed, telling Cortelyou he was in no pain. "Let no exaggerated reports reach Mrs. McKinley," he told his secretary. Then he said it again, to be sure.

The fanning went on. Six men, not knowing what else to do, joined in.

Outside the Temple of Music, fairgoers, drawn to the sounds of calamity, began to cluster and agitate. "Kill him!" someone shouted, and another man, in the unfortunate parlance of the day, yelled, "Lynch him!" An electric ambulance pushed through the group and, within minutes, attendants had McKinley on a stretcher. As they emerged from the building, Randolph Keim put a hat over the president's face, shielding it from spectators. There was no mistaking the big man, though, and some onlookers started to sob.

The trip to the Exposition Hospital was fast. Keim, who held onto the back of the ambulance, also noted that it was exceptionally smooth, with no jostling and little vibration. The journalist attributed this to good rubber tires, and also to the asphalt. In almost any other circumstance, Buffalo would have been proud.[1]

Surgeon Matthew Mann, a specialist in gynecology, was at a barbershop, mid-trim, when a courier ran in and told him to hurry to the Exposition Hospital. He arrived at 5:10 p.m. to find general surgeon Herman Mynter already there. Famed Dr. Roswell Park, the Exposition's chief medical officer and an expert on gunshot wounds, was in Niagara Falls finishing neck surgery on a patient. It would be a while before he could get back to Buffalo.

Mann and Mynter, along with McKinley's faithful Dr. Rixey,

decided to move ahead. One of the president's wounds, they could see, was superficial—a bullet had fallen from his clothes—but the other was deep. It had entered the president's abdomen, and was lodged—somewhere. Nurses shaved McKinley's belly, wiped it with alcohol and green soap, and administered ether. After nine minutes, the president was unconscious, and the surgeons, with Mann taking the lead, went to work.

Escaping gas soon told the physicians that a hollow organ had been perforated, and further searching showed that the bullet had gone into and exited the stomach. They sewed up these holes. Where, though, was the bullet? "The greatest difficulty," commented Mann later, "was the great size of President McKinley's abdomen and the amount of fat present." Working in his belly, he said, was like "working at the bottom of a deep hole."

Nevertheless, Mann investigated, and, in one of surgery's most elegant euphemisms, he "introduced his arm," and probed. The search began to have an effect on the patient—he was showing signs of shock—and they knew they faced a choice: continue to "eviscerate" the president (there was no nice word for that one) or stop altogether.

They got out. They cleaned the bullet track, irrigated the site with hot salt solution, and sutured the wound with silkworm and catgut. Just as the men were finishing, the door opened and Dr. Park appeared. The square-faced Park, known for both his cool reserve and his professional talent, had had to wait for trains at the Niagara Falls station. While the surgeons closed the incisions, Park took a back seat, but then he called the surgical team into his office. Whatever they felt, he said, and however they might not concur, they were now to present a "united appearance." Keenly aware of recent history, he urged them to "avoid all the discussions and differences of opinion which had so conspicuously marked the conduct of President Garfield's case." Park knew that the reputation of Buffalo medicine, not to mention the good name of the city itself, rested on decorum. That and a good outcome.[2]

Fairgoers wait near the Exposition Hospital as
President McKinley undergoes emergency surgery.

Roswell Park did not actually agree with the treatment he witnessed at the hospital. A decade later, from the safe distance of time
and with the help of hindsight, he confessed he had seen things that
bothered him. He didn't like the way that Dr. Mann used one of
his surgical instruments to "rap" Dr. Mynter over the hand to get
him out of the way. It had not been the time, Park felt, for Mann's
"petulance." Park also thought that surgical caps should have been
used. Dr. Mynter hadn't worn a cap, and the other two surgeons
watched "drops of perspiration fall into the wound from Dr. Mynter's forehead." Park explained that this was not due to nervousness:
"Mynter . . . always perspired freely when doing any of this work
in a warm room."

There were other omissions. Why had the wound not been drained
more aggressively? Park himself had assented to the lack of a drain,
but it had been a "hurried" decision. And it was disturbing that the

surgeons had not used the right retractors or needles. After McKinley's operation, Park discovered a complete, unused surgical case sitting in his office.

And then there was the problem with the light. It would become the great irony: The fair with the grand Illumination, the place where electricity was more plentiful than almost anywhere else on earth, the Exposition applauded by Tesla and Edison both, and they couldn't get a decent light bulb.

The operating room, darkened by an awning, allowed the afternoon sun into the room, and this was just enough to bother eyes but not sufficient to light the site. It was Dr. Rixey who came to the rescue with a hand mirror, and he angled the slanting rays into position. Finally, well into the procedure, an electric light arrived—eight watts. As people commented later, it would have barely brightened a Christmas tree.

Why, Roswell Park also wondered, had they not waited for him? He would have found light and instruments and would have insisted on different ways of doing things. The surgeons told him that they couldn't wait; the president was weakening. Park disagreed, and wondered whether there was more going on than medical decision-making. Maybe the rush to surgery had been shaped by something else. Maybe it was "jealousy." Others would have the benefit of hindsight, too, of course, and would contend that if Dr. Park had been there, it would not have made a whit of difference.[3]

The sun was close to setting when President McKinley was loaded back into an ambulance for a ride to the chosen place of recovery—John Milburn's house. The procession, made up of police, guards, and men on foot and on bicycles, moved slowly in the fading light. There was no rush now, only a need to keep the patient as still as possible and free from pain. McKinley was beginning to show signs of waking up.

As the ambulance approached the Lincoln Parkway entrance, Rainbow City, as it always did at dusk, began to come to life. Its buildings, budded with lights, began to glow red, and then brightened to a cheery yellow. For about thirty seconds, the illumination expanded. Then, just as the procession reached the gate, the lights cut off, and dark descended like an untethered and ponderous curtain.

On a night like this, someone had decreed, there should be no beauty.[4]

The news of the shooting spread around Buffalo in fits and starts—in bold headlines, in shouts from street corners, and in urgent messages that shattered nerves and dashed plans. Pan-American director Frank Baird had plans like these. Following the hot day at the falls, and after seeing the president off for his reception, Baird had stopped at the Exposition's Log Cabin for a drink and a talk. He wanted to hear Colonel Webb Hayes, recently returned from the Philippines, tell stories of army life at the front. It would be a short presentation, Baird hoped, because he had a big evening in the works: George Williams, the treasurer of the Exposition, was hosting a formal dinner for the McKinleys. In fact, Williams had been preparing for days. He had ordered his house on Delaware Avenue, designed by the famous Stanford White, dusted and polished like a piece of fine furniture. He had sent servants up to the attic to lower the big gold-plated chandelier so that it could be fitted with candles. And that morning his wife had delivered to each of the dignitaries—Exposition directors, cabinet officers, and of course the McKinley group—a little pasteboard box containing a boutonniere.

They had not been at the Log Cabin long, Baird remembered, when one of the party pulled out his watch and realized the lateness of the hour. It was almost four o'clock. Time to secure a carriage, drive home, and, with their wives, do all the bathing and powdering and boutonniere-fastening that the state occasion required.

Baird, whose house was the closest to the grounds, shared a ride and got out first. He did not watch the carriage as it moved down Delaware Avenue. Had he done so, he would have noticed that two blocks farther along, someone hailed the carriage and ran, frantically, up to the driver.

It was his wife, who had been out calling on some Ohio acquaintances, who flew inside with the news. "The president has been shot," she cried. Baird did not believe it. Running to his carriage, he ordered the driver to put the horses into a gallop. Racing up the avenue and back through the Exposition gate, he shouted, "Is it true?"

The governor was in a boat on the Erie Canal, inspecting the waterway, when he heard.[5]

And Vice President Roosevelt was on an island in Lake Champlain, three hundred miles to the east, when the news came by telephone.

Roosevelt had been in his element, banqueting outdoors with hunters, fishermen, and fellow politicians. He was the guest of honor at a meeting of the Vermont Fish and Game League at the Isle La Motte estate of former lieutenant governor Nelson Fisk. The sportsmen had dined under a tent and listened to a rousing rendition of "To Arms" by the St. Albans Glee Club. Then toasts had been proffered, and Roosevelt had addressed the men.

"I am," he told the group, "interested in all furred, finned and feathered inhabitants of the woods and waters." Hunting and fishing, he said, cured boredom, and wilderness itself turned a profit. He used the example of a deer. "A dead deer is worth a few dollars, but a live deer is worth much more. It is a bait for city sportsmen. They do not always hit the deer and they . . . leave a hundred times the worth of the deer in money."

It was a short speech, and afterward Roosevelt excused himself and walked into the Fisk house to prepare for the evening reception.

Somewhere in the interior, a phone rang.

It was for the guest of honor. Roosevelt was located and given the earpiece.

He took it, listened for a minute, and then dropped it. He put his hands up to his temples.

"My God," he said.

Minutes later, Senator Redfield Proctor walked to the stone portico of the house to deliver the bad news to the group. "Gentlemen," he said, "it is my sad duty to announce that word has just been received by telephone—I trust that it may prove false—that . . ."

Nearby, a steamboat let out a shrill whistle. What was it that Proctor was saying? It was hard to make out his words. Then the boat moved away, and Proctor finished. In a single breath, the crowd groaned. Some broke into tears.

There was no shortage of men to help move Roosevelt off the island and toward Buffalo. Dr. William Seward Webb offered his steam yacht *Elfrida* to take Roosevelt back to Burlington in order to catch a special train south. Other men ordered private railway cars. They would use all the power and influence they could muster to get the vice president to McKinley's bedside and to the site of the dastardly act.[6]

II
THE HOUSE NEAR THE CORNER

Running through Buffalo's heart, Delaware Avenue carried the crucial work of the moment. From its foot close to the lake, the avenue moved north and south through the city, past the house of Ansley Wilcox, where Vice President Roosevelt arrived in a hurry at midday on Saturday, September 7. It then ran a mile uptown, past grand edifices, trimmed lawns, and draping trees, to the Milburn house, where the president lay on his back in a hospital bed. A mile farther on, it flanked the Exposition. The fair was still open, going through the motions of existence.

It was the Milburn house, near the corner of Delaware Avenue and West Ferry Street, where most people wanted to be. They wanted to know the latest, to see who came and went, and to assess their expressions and steps. But they couldn't. A block from the house in every direction, police put up barricades, and the crowds that gathered, the trucks and the carriages that slowed, could barely get a view of the house's ivy-covered garret.

The first hours, then the first days would be the most trying, doctors said. If McKinley made it through those, the dangers of blood poisoning diminished. The physicians promised to do their best to let everyone know, all the time, how things stood. And they would not worry about delicacy. If the American public had previously appreciated William McKinley distantly, as a dignified man in a waistcoat and a shimmering top hat, they would now get to know him on a more personal level. They would be apprised of his pulse rate and his urinary output. They would learn how he got his nourishment: hot water by mouth, and whiskey and egg through the rectum.

The doctors ordered silence. Conversations with the patient—even for Ida McKinley—had to be brief. Footsteps had to be muffled. Telegraph machines, whose tapping could be heard in the house, were sent farther away, and trucks were forbidden. Men brought in ice and groceries by hand, carrying loads on their backs. Army soldiers who trod back and forth were ordered to use the grass, to dampen the noise of their boots. Nobody could do anything about the bell of the streetcar, though, and, blocks away, its clang reminded listeners that, for some people, life went on.[7]

Outside the Milburn residence, under an old army tent, a cadre of reporters put up tables and chairs and waited and watched. They took their clues from the house itself. They thought they knew the windows—which ones belonged to Cortelyou, which belonged to Mrs. McKinley. They thought they knew the servants' quarters, too, and kept their eyes on those for any sign that someone was up too long or too often. And they tried to imagine the president's room, up on the second floor. They were told that John Milburn's boys had slept there once, and that McKinley lay in a white iron bed. The shades were drawn most of the time, and the room was cooled by fans.

For the newsmen who had served in Cuba and the Philippines, and for the old-timers who remembered encampments in the South, the emergency scene at the street corner had to have roused flashbacks to army life. Sentries from the Fourteenth US Infantry, with their rifles and bayonets, walked an incessant square around the Milburn house. Bodies, too, lay scattered here and there. These must have given some

Secretary George Cortelyou delivers medical
bulletins to reporters near the Milburn house.

reporters a jolt, until they realized the recumbent forms were just exhausted newsboys, curled up against the cold.

The weather worsened as the weekend progressed, and journalists did what they could to stay warm in the wind and rain. They brought in a coal stove to take the chill out of the air, and, when that did not work, they tried to find some heat by smoking. At night, kerosene lamps and an old streetcar headlight brightened the tent enough to allow them to work. Many of the men had not changed their clothes since the shooting.

The first bulletins on the seventh and eighth of September told of a man injured but holding his own. Twenty-four hours after surgery, McKinley's pulse was a high 130 and his temperature had risen to 103°F, but doctors maintained there was "no change for the worse." The medical team, led by Roswell Park, also said they would know soon whether the bullet could stay in the president's back muscles, or whether it had to be removed. There were X-ray machines in Buffalo, but McKinley was a big man and required different equipment. The Edison Laboratories were sending an apparatus up from New Jersey.

In the morning after the shooting, the president asked for his wife, who had barely left her own bedroom. She came in, and they held hands. He told her he was not in much pain and that the night had been restful. "You know," he added, "you must bear up well. That is the best for both of us."

The physicians reported that, after this visit, his pulse fell and his respiration calmed.[8]

While President McKinley took nourishment from enemas, and quenched his thirst with teaspoons of water, Fred Nieman, swollen and bruised but otherwise uninjured, ate heartily at the city jail. His cell, deep below ground, was known as the "dungeon." No sun reached the dungeon, only the fitful yellow of gaslight, and the prisoner was left to the stimulation of his own mind. For once, he did not have newspapers.

But he did have food: roast pork and lamb, fried ham and eggs, boiled potatoes, succotash, sliced tomatoes, homemade bread and butter, coffee. "That man is a d——d glutton," commented a police captain. People didn't like hearing about his ample diet. Honest laboring men would be lucky to have that food once a year, one man complained. But the jailers weren't simply being generous. The healthier the prisoner was, the better able he would be to stand trial. And to be convicted. And executed.

When Nieman wasn't eating or sleeping soundly, guards reported that he stared. He also laughed strangely and seemed oddly unaware of the magnitude of his crime. To the questions that were put to him by police and lawyers, he offered truth and lies. One truth was that his given name was not Fred Nieman but Leon Czolgosz. A Buffalo paper took pains to explain that the surname was pronounced "Shull-guss."

Czolgosz told officials that he had tried to kill the president "because I done my duty." One man had so much power, he said, and others had so little. A detective asked whether he was an anarchist. "Yes, sir," he replied. He said he had been studying anarchism for a while, listening to speakers, reading papers, meeting with believers.[9]

Out of everything that interrogators soon learned about Czolgosz—including the struggles of his immigrant family, his work as a laborer, his unemployment, and his illnesses—it was his anarchism that drowned out everything else. Who cared what had made the man go in this dreadful direction? It was this unspeakable belief and the act that followed that mattered. The hysteria of the 1886 Chicago Haymarket affair suddenly seemed fresh. So did the anarchist assassinations of the French president in 1894 and the Italian king in 1900.

By admitting his devotion to anarchism, Czolgosz also helped his prosecutors. They would worry less now about an insanity defense and could avoid the sickening charade that the assassin Charles Guiteau had perpetrated with James Garfield. Radical anarchism—the desire to create a stateless world through terror and violence—was an abominable belief, but people did not have to be certifiably insane to adhere to it. It was evil, but—this was a fine distinction—not madly evil.

As soon as police officers extracted the word *anarchism* from the prisoner and handed it to the press, fury ignited. Across the country, crowds vandalized anarchist newspapers and printing presses and mobs threatened anarchists in their homes. In Tacoma, Washington, a group formed to "annihilate" the political radicals, and in Pennsylvania, two dozen armed men attacked an anarchist community and forced twenty-five families from their homes. In New York City, police put all known anarchists under surveillance. "The purpose of this," announced the commissioner, was "to make life so disagreeable for anarchists in New York City that they will move out of it."

For all their outcries against the lawlessness of anarchy, some attackers ignored the legal process themselves. "Lynch him," men yelled at a target in New York when they gathered to mete out vigilante justice. In Cleveland, a group of African American veterans organized to denounce both anarchism *and* lynching. Other black citizens strategically reminded the public of their longstanding loyalty to the nation.

Rage burned hottest where Czolgosz had contacts: Buffalo, Cleveland, Chicago. In Buffalo, Chief of Police William Bull flung a wide net over the city, particularly the immigrant East Side. Police in Ohio swept through Cleveland neighborhoods, locking one "foreign looking man" in a "sweatbox" overnight. Chicago police arrested thirteen alleged anarchists, including Czolgosz's would-be mentor, Emma Goldman. When asked about his association with Goldman, Czolgosz denied her connection with the shooting but couldn't resist commenting that "she is a good woman, a friend of the poor man and an enemy of the plutocrat and the monarch."

It didn't matter that some anarchists eschewed violence, or that some "anarchists" were really socialists. It was not the time for anyone to critique the government of the United States. On Sunday, September 8, Buffalo police stopped a meeting of the Socialist Labor Party because they were afraid someone might say something negative about the president. Boris Reinstein, one of the speakers, explained that the "object of the meeting called for tonight was to . . . condemn emphatically the outrage of last Friday." Reinstein also noted that, as a socialist, he was "opposed to the capitalistic arrangement of society, but our methods are to accomplish reforms by the use of the ballot." The police chief canceled the meeting anyway.[10]

III
THE VORTEX

While the country wrestled with its wrath, the Pan-American fair sputtered on.

"We sincerely hope," commented a tone-deaf official, "that the feeling will not get over the country that a gloom has been thrown over the Exposition." It was, he added, the president's desire that the fair "will be as great a success as we have always hoped." The director of concessions, Frederick Townsend, echoed the sentiment with enthusiastic, if inapt, words. The Midway, he announced on Saturday, September seventh, will be open. "Everything will run full blast," he declared.

While Exposition directors, concessionaires, and local labor organizations issued resolutions expressing sympathy and concern, committees began to cancel events. New Jersey Day was put off and a committee of Poles canceled Polish Day.

Some determined visitors did try to enjoy the Exposition. Twenty-five thousand of them, in fact, entered the gates on Sunday, September eighth. But the mood was somber, and guests seemed saddled with worry. An elderly couple confessed that they fixed their eyes on the flag at the end of the Esplanade, fearful that they would see it lowered. "Thank God it is there yet," the old man said. "I hope it will never fly at half mast."

Eager to tap into the swelling emotion, exhibitors pulled out every photograph they could find of McKinley and charged customers double and triple. People paid. At the scene of the crime, visitors tried to determine exactly where the president had been shot. What remained of the reception had been cleared away, but the Temple of Music opened for organ concerts on September 7, and opportunistic relic hunters went to work. Assuming that McKinley had stood on the stage in the rotunda, they began taking pieces of wood by cutting or tearing the stage apart. A carpenter had to be called in to make repairs.

Over on the Midway, the Animal King put on a brave face. Out of respect for the president, he announced, he would postpone his parade of the lion-wedding cage. He didn't mention that he couldn't have staged his parade anyway—police had blocked the streets.

Bostock's press agent, Captain Maitland, also shifted gears. Instead of highlighting rambunctious or bloodthirsty animals, he found a more dignified focus—the Animal King himself. While other publicists delivered tributes to the injured president, Maitland handed press offices a looming portrait of Bostock. Two days after the shooting, readers were also offered a complete biography of the menagerist, including details of the accolades awarded him by the Prince and Princess of Wales.[11]

Beyond Buffalo's perimeter of woe, in Niagara Falls, entertainers continued to market their acts. In fact, one of them, Maud Willard, Martha Wagenfuhrer's friend, was ready to perform. Oddly enough, twenty-five-year-old Willard hailed from Canton, Ohio, just as McKinley did. Still odder, perhaps, was the revelation that she was no stranger to the matter of presidential assailants: Once upon a time, she had worked in a theater company with the brother of an assassin, Edwin Booth.

Willard hoped to work her way out of hand-to-mouth living and to support her ailing mother. All she had to do was what her friend Martha had accomplished the day before: perform a barrel act in the wild Niagara River. Martha now had a marketable new name: "Queen of the Rapids." How bad could it be?

Willard's barrel ride would actually be a double act with an experienced stuntman named Carlisle Graham, who had ridden the rapids five times. He would lend Maud his own well-tested barrel. But the

Carlisle Graham and Maud Willard pose before their stunt.

plan was complicated. Maud would ride the barrel through the treacherous rapids below the falls, and then, if all went well, enter the Whirlpool. Graham would jump into the river opposite the Whirlpool and
swim with her barrel toward Lake Ontario. Her stunt and his long
swim would be record-breaking. A moving-picture company would
record the feat.

If there was such a thing as a good day for barrel riding, September
seventh claimed it. The weather was hot and rainless. At 3:30 p.m.,
Maud Willard made her way to the dock of the *Maid of the Mist* tour
boat and announced she was ready to go. In honor of the stricken president, her fellow Ohioan, she carried an American flag. And in honor
of nobody in particular, but just because she could, she took along a
small brown-and-white fox terrier. An hour later, she hoped to have
survived the Whirlpool Rapids, boosted her income, and become a
champion of the fallen American leader.

Having been nearly killed by her own barrel run, Martha Wagenfuhrer had done her best to urge Maud to drop the idea. Others
wondered about taking the dog. What was the point? There was
only one small airhole in the barrel. She was told to leave her dog
ashore.

Maud didn't listen.

At ten minutes to four, in view of hundreds of spectators clustered along banks and jammed overhead on the steel-arch bridge, a
small boat towed Maud's barrel to the center of the river and cut it
free. Within seconds the barrel was thrown about by the rapids, but
it remained upright and moved fast. In less than eight minutes, it had
traveled into the Whirlpool.

Carlisle Graham, who had witnessed the launching of the barrel,
hopped on a private railway car to keep up with it. He saw Maud

enter the Whirlpool and watched as she made one wide revolution and edged closer to the Canadian bank. The barrel took longer to move around the Whirlpool than he had expected, but Maud would be hooked and brought ashore after one more circle. He was sure of it.

In the meantime, Graham would do the river trip alone. Donning a cork life preserver and an inflated neck tube, he jumped into the rapids below the Whirlpool, powered through the currents, and entered the river's calmer stretch, a mile from Lewiston, New York, his destination.

Graham completed his swim and took time to rest in Lewiston and savor his success. Close to 6 p.m., he headed back to the falls on the Gorge Railway. It was a pleasant evening along the lush, tree-banked river, and Graham chatted with friends and admiring strangers. If anxiety over Maud shadowed his pleasure, he did not acknowledge it. He had not seen what others had witnessed—that, at 5 p.m., after Maud had made more than thirty revolutions around the Whirlpool, her barrel had been sucked down by an undercurrent and had disappeared for several minutes.

It wasn't until Graham's railroad car approached the Whirlpool that he realized what had happened. Even in the slanting light, he knew that the dark shape in the water was her barrel—his barrel—and that it was trapped in the spiral. As soon as the train got close enough, Graham jumped off, hurried to the Canadian shore, and dove in. He was strong, but not strong enough to counter the current, and, along with everybody else, he could do nothing but wait. The barrel listed now—its contents had shifted.

At around 8 p.m., four hours after Willard had begun her trip, men brought in a searchlight car from the Gorge railway and focused its beam on the water. In what was becoming a macabre ritual on the riverbanks, they lit bonfires. Half an hour more, and a man spotted a piece of wood drifting closer to shore. The current had changed. Then the barrel bobbed into sight. Two local brothers swam to the cask and pulled it to the rocky bank. Maud Willard and her terrier had been encased nearly five hours.

Some witnesses recalled that one of the men who pulled Maud out of the barrel, Captain Billy Johnson, thought there was hope left, and tried to resuscitate her. Johnson, a former lifeguard, struggled for more than an hour. Another person ran for a doctor. The physician was not optimistic, and, at about 10 p.m., he pronounced her dead. She hadn't died easily, he concluded. Her lips were blue, her body was bruised, and "the facial expression denoted much suffering."

Graham paced from rock to rock to rock in anguish. He spoke of his last words to her. "Don't open the cover," he had told her. And she never had. There wasn't even a "pailful" of water in the barrel when it was opened.

On other matters, though, Maud hadn't taken anybody's advice. She had carried a dog with her, for God's sake. And when the men broke open the top of the cask and looked in, there he was. He was sitting on Maud's face, his nose to the airhole.

The good news? He was very much alive.

It was bad enough that her daughter had died in such a gruesome way, but Willard's mother was horrified by descriptions of the incident. Reporters claimed that her daughter's body had laid out on a rock for hours, and that the ghoulish glow of the firelight had created a scene "not unlike Dante's Inferno." A Toronto paper claimed that Maud's friends had pulled her corpse up the embankment by her hair and ankles and that they had been so drunk that one or two of them climbed into Maud's coffin and passed out.

It was all too much. Two weeks later Mrs. Willard, like her daughter, was dead.[12]

Stories of Maud Willard's suffocation circulated widely. Michigan papers, including the Bay City press, reported the failed stunt, so Annie Edson Taylor, would-be barrel rider, certainly read about it. In public, at least, she barely blinked an eye. And why would she have? She believed she stood worlds apart from the likes of Maud Willard. Not only had she studied the matter for months, but she was a woman of dignity, education, and, above all, class.

At the same time, Taylor would have to admit that progress in Michigan was slow. The cooperage had made a mistake with the barrel cover—it was not big enough. There also was the matter of a manager. The first three men Annie approached refused. Finally, she located a Mr. Russell, the "best advertiser" in the city, who said he would take her on. Russell had even sent her a typewritten letter—a sure sign of reliability.

Then there was the problem of money. After a wealthy acquaintance turned down Annie's request for assistance, she determined to fund the project herself. She mortgaged everything she had: her sheets and pillows, her tablecloths and linens. All she had left in the world was a trunk full of her last possessions—and a nicely outfitted, human-sized cask.

Annie Taylor set her sights on the end of September. In the fullness and warmth of early autumn, Exposition crowds would certainly surge to the falls to see her. And after she made her miraculous trip, they would pay to see her in person. The wonderful Mr. Russell—Tussie, as he was known—would arrange the appearances: Niagara Falls, Buffalo, and, above all, the Pan-American Exposition.[13]

IV
GLORY DAYS

There was no good news out of Niagara Falls on September 7, but upstream in Buffalo, the other resident of Canton, Ohio, was holding his own. By the evening of September 8, in fact, people emerging from the Milburn house were almost beginning to smile. McKinley's aides reported no signs of impending fatality: no indication of blood poisoning, infection, or obvious abscesses. Senator Mark Hanna, the president's good friend from Ohio, announced that the bullet, wherever it was, could become incrusted in forty-eight hours, and then be left alone, forever.

The doctors were more cautious. On Monday, Dr. Matthew Mann, who had performed the surgery, admitted to reporters that he had "known cases to go well for ten days and then change for the worse." The celebrated abdominal surgeon Dr. Charles McBurney, who had arrived from New York City on Sunday to join the medical team, claimed that the president would not be out of the woods for another week. McKinley, he said, was "far more confident than anyone around him."

But every hour that went by, then every half day, seemed to move the victim closer to recovery.

By Monday, even the surgeons were beginning to show relief. Asked about the special X-ray machine that had been hauled up from Edison Labs, McBurney said it wasn't needed now. The only point of that, said the surgeon, would be to take out the bullet, and why would anyone do that when it wasn't causing any trouble?

Newsboys could not be contained. "Extra!" they shouted. "The president will live." Editors echoed them. "Crisis About Over," claimed the *Commercial* on September 9. "President McKinley's condition is now so favorable that it has dispelled almost the last shade of doubt and apprehension." The next day, the paper was even more emphatic: "Now the President's recovery is certain." The searchlight at the Electric Tower sent an uplifting signal message to the tower at Niagara Falls: "The country rejoices at the rapid recovery of President McKinley."

Reporters who wondered whether the doctors were just telling the public what it wanted to hear were given an emphatic *no*. Vice President Roosevelt, still in town, was impatient with this sort of suspicion. "I want to say that the official bulletins are scrupulous understatements of the hopefulness of the situation," he insisted, pounding his fist into his hand. Senator Hanna concurred. No physician wants to be overly optimistic, he insisted. "Their reputation is at stake and they have a right to guard it."

When journalists carried on about the president's high pulse, they were told to stop their fretting: It was probably due to secondary shock. McKinley's heart rate—and the McKinley family doctor, Rixey, confirmed this—had a "trick of quickening" anyway.[14]

Still, skeptics recalled another president who had been shot, twenty years earlier. Wasn't James Garfield's case a caution to everyone: early optimism, and then a horrendous crash to fatality? They were assured the cases couldn't be more different. The agonizing months waiting for Garfield to recover lacked one critical element: the president's own desire to live. One commentator insisted that there stood over Garfield a strong sense of impending death. In fact, the insider argued, "Garfield believed from the beginning that he would never recover. He spoke words of cheer and comfort to those about him, for the dear sake of family and friends, but his physicians . . . knew that the man's heart was not in the struggle."

All this was reinforced by the suggestion that Garfield's spirit had been broken by being shot in the back, unprotected and unsuspecting.

McKinley, shot facing his enemy, was a soldier, and was determined to get well—for his wife's sake if nothing else. Dr. McBurney summed up the opinion that there could be no comparison between Garfield and McKinley: Garfield's wound was "an extremely unfortunate one, hard to get at and difficult to handle." By contrast, McKinley's wound was "a fortunate one, a lucky one."

By Tuesday afternoon, September 10, the cloud of disquiet that hung over Buffalo began to lift. The expressions of those who emerged from the Milburn house had broadened into wide smiles. Senator Hanna, reported one observer, "bubbled over with joy." Doctors were hardly less thrilled. "I feel just like hollering," said one. And the president? He "was in a fine mood," too, and wanted to chat with everyone, about anything.

Mrs. McKinley, who had been subdued with medication and, except for fleeting conversations with her husband, had kept to her room, celebrated the general optimism by taking a carriage ride. The next day, she took another.[15]

The biggest indication that Buffalo itself was coming back to life was that the city began to debate matters other than medical prognoses. There was the question of women's right to vote, for example. Suffragists, including celebrities such as Susan B. Anthony and Carrie Chapman Catt, arrived in the city on September 9 and 10 for the meeting of the National Woman Suffrage Association. Mayor Diehl welcomed the women at the City Convention Hall, and the activists launched into a spirited series of lectures and discussions about how the franchise would "create happier homes," "advance the cause of peace," "purify politics," and "develop the higher manhood of men."

Vice President Roosevelt, among other high-level politicians in Buffalo at the time, steered clear of the suffragists. As governor in 1898, Roosevelt had expressed lukewarm support for women's suffrage, but he also believed that white women should focus on "duties" instead of rights. It was a white woman's duty, he would later assert, to be the "helpmate, the housewife, and mother," and she should focus on bearing children—lots of them.

While the suffragists rallied downtown, Roosevelt ambled about other parts of the city. He walked between the Wilcox and Milburn houses on Delaware Avenue and took a trip to the zoo. As for riding in a protected vehicle, or having more than a couple of security men near him? "Pshaw!" he said. A laborer stopped the unguarded Roosevelt as he strode about and asked him if he wasn't afraid to be approached. "No sir," Roosevelt snapped. "And I hope no official of this country ever will be afraid. You men are our protection and the foul deed done . . . will only make you the more vigorous in your protection of the lives of those whom you elect to office."

By the beginning of the week, confident officials and medical advisers began to exit the city. Dr. McBurney and Vice President Roosevelt left Buffalo on Tuesday night, September 10. Reporters, who seemed strangely obsessed with the past, asked Roosevelt once again about Garfield. He had seemed so well, they reminded the vice president. "Ah, but you forget twenty years of modern surgery, of progress," Roosevelt replied, and got on his train. Charles McBurney, who planned to visit Niagara Falls before heading home to New York City, was not only certain but expansive. "Gentlemen," he was heard to say, the president's recovery "is the event of the century."[16] The papers reported his words the next day, September 11.

V
"BIG JIM"

Now that the president was on the mend, there were heroes to thank. There were the doctors, of course, and there were the Secret Service men and detectives who had acted so quickly and selflessly. And there was Jim Parker. Hardly a day had passed after the shooting before the waiter was the talk of the town. His sudden fame had even added inches to his height. Forty-four-year-old James Benjamin Parker became known as "Big Jim," and he had stretched to six and a half feet tall. He became the fair's hero, its "tawny lion."

Parker, according to some observers, took his new fame in stride. "I happened to be in a position where I could aid in the capture of the man," Parker was heard to say. "I do not think that the American people would like me to make capital out of the unfortunate circumstance. I am no freak, anyway. I do not want to be exhibited in all kinds of shows. I am glad that I was able to be of service to the country."

Parker's role in subduing the shooter was splashed across the front pages of the nation's papers, and sparked widespread celebrations. In Savannah, Georgia, where Parker had lived and worked as a constable, the black community was delighted. They remembered Parker's brave work in that metropolis and hoped that he might be rewarded with a job in Washington. In Syracuse, twenty-five percent of the receipts from a performance at the Grand Opera House was scheduled to go to a fund for the man "who saved President McKinley's life." And in Buffalo, Jim Parker not only received offers to appear at the Exposition but also was credited with singlehandedly breaking down the barriers of segregation. The president and secretary of the Don't Knock Society, which rarely accepted black members, decided to invite the courageous Parker to join them.[17]

It is unclear whether the meteoric rise of Jim Parker did anything to mitigate the anger felt by Buffalo's African American community toward the Pan-American Exposition. Since 1893, when Ida B. Wells and Frederick Douglass lambasted the directors of Chicago's White City for race discrimination, and protested the absence of an African American exhibit, United States fair directors had had an uneven history of recognizing black achievements. Atlanta's exposition had constructed a Negro Building for exhibits and conferences in 1895, but other fair directors, including Buffalo's, were more interested in showcasing African villages—with cannibals, wild dances, and witches—than progressive developments in the contemporary diaspora.

When the Pan-American's Board of Managers took shape as an all-white body, and then said yes to Darkest Africa and yes to the Old Plantation, with its living exhibit of slavery's "good old days," Buffalo's progressive black community—especially its women—spoke out. Rallying at the Michigan Street Baptist Church in November 1900, the Phyllis Wheatley Club of Colored Women protested the exclusion of people of color from the Board of Managers of the Exposition. Led by Mary Talbert, an Oberlin graduate and community organizer, the group also argued that the Exposition needed to highlight African American progress. And they had just the show: W. E. B. Du Bois's prizewinning Negro Exhibit from the 1900 Paris exposition.

Mary Talbert was tireless. Before the Pan-American Exposition opened in the spring of 1901, she staged a fundraising show that featured African American talent. She also helped host a conference for the National Association of Colored Women during the fair season, bringing to the city some of the most celebrated and enterprising women in America.

And she got the Negro Exhibit to the fair. Boosted by an appeal from the exhibit's curator, Thomas Calloway, Talbert and other clubwomen saw the award-winning show installed in the Manufactures and Liberal Arts Building. Through the exhibit's charts, graphs, publications, and photographs, visitors learned about the progress of black Americans since Emancipation. They read that the literacy rate of African Americans was now greater than that of Romanians and Russians, and they viewed inventions and patents. They peered into dioramas to follow a family as it moved forward from the 1860s. Nine models, lit by electric lights, showed the group emerging from the shadows. "Behind them," explained the curator, "are woods representing the darkness of slavery, and before them is a winding path leading into an unknown future." The next model pictured the family building a small house as well as children clustered under a tree, learning their lessons. The final scenes revealed the family's son, grown up, teaching his own classes in a "neat white schoolhouse with glass windows and a brick chimney" built by the community. It was a story of challenge and promise.[18]

Mary Talbert.

Mary Talbert and others put passion and effort into celebrating African American accomplishments at the Exposition, but they climbed a steep, often insurmountable hill of public opinion. While African American newspapers lauded the show, hoping that it would dispel the "gloom" directed at the black community, the Negro Exhibit might as well have been off in the center of Lake Erie for all the attention white publicists gave it during the Exposition season. Most guidebooks, including authorized catalogues, ignored it. The local press, with the exception of the *Express*, turned its back on the show. The *Express* tried to be generous, explaining that an exhibit "more complete, valuable, and far-reaching in its effects could scarcely be imagined." Then it qualified its enthusiasm, asserting that it would mostly be of interest to black visitors.

The *New York Times* acknowledged the exhibit, barely. In commenting on its display of literary achievements, including works by Frederick Douglass and W. E. B. Du Bois, the paper took pains to be unflattering: "We may as well be entirely frank in the appraisal. Much of it is rubbish. None of it is very great."

African American visitors who saw the Du Bois display were not interviewed by the white press, at least in Buffalo, and they left behind few, if any, printed accounts. Almost a century passed before local historians discovered a pamphlet describing the exhibit and realized that it had even been installed at the Exposition.[19]

While white fairgoers likely bypassed the Negro Exhibit, they did attend other performances about people of color on the Midway. Menagerists elicited big laughs from visitors by showing how their well-trained animals outsmarted black performers. At the end of June, an African American performer from Georgia became the "pupil" of a horse. "The darky is learning to read and write," explained a reporter. "His teacher is the black horse, Bonner, who not only reads and writes, but does sums in arithmetic." Not to be outdone, Frank Bostock revealed that his chimpanzee, Esau, got on well with an "Old

Plantation Negro." The two were publicly introduced to each other in the monkey's "private apartments." They shook hands and joked together. "Ha-ha-a-a-a-a-a-a-!" laughed the black man. Esau laughed in return. "Waugh! Waugh! Waugh!" he said.

Teacher Mabel Barnes never saw (or, if she saw, she never acknowledged) the Du Bois exhibit, but she was a keen fan of Darkest Africa and the Old Plantation. She and Abby Hale visited Darkest Africa in the summer, lured into the stockade by a gyrating "real pygmy" in a cotton loincloth. There she encountered "village life" performed in bamboo huts by nearly one hundred West and Central Africans. Surrounded by shrieking parrots and monkeys, Mabel and Abby watched African weavers and gold and ivory workers and became wide-eyed at the sight of drums, spears, and idols. They listened to stories of human sacrifice and polygamy. Mabel took notes on "specimens of pygmies and cannibals." Many of them, she observed, had never before left home.

The women were not shy about their curiosity. Of all the demonstrations at the Darkest Africa concession, the most appealing, said Mabel, were "the natives themselves." Their bodies, she confessed, "are slender, strong, and clean." She liked their white teeth, too. But most of all she admired their skin. It gleamed with palm oil, the schoolteacher said, and looked like satin. She admitted it was "so clean and smooth that it almost tempted one to lay hands upon it."

For their part, the performers reminded the women that they were more than specimens. "Miss Hale and I were making some comment upon [the skin]," explained Mabel, "not realizing that the owner of the skin, who, moreover, was of rather an intelligent appearance, could understand English." For a split second, the two parties closed the distance between them and spoke. The women also discovered an African ivory worker returning visitors' stares. Using his carving skills, he was busy reproducing a "Midway type," an American woman, with "excellent fidelity." Mabel was impressed—sort of. "As simple as they are," she commented, "these natives are also quick at observing the customs of the world."[20]

Behind the doors of a columned antebellum façade at the Old Plantation, Mabel and Abby found more to learn about dark-skinned people. E. S. Dundy's concession featured antebellum life in the American South, and slavery never saw such an endorsement. Slave families sang and danced and laughed as they picked real cotton. Of all the performers, nobody made the case for happy days in the Old South better than the man who had laughed with Bostock's chimpanzee. Laughing Ben, diminutive and gray-haired, was an honest-to-goodness ex-slave. Opening his mouth wide—wide enough that visitors could count his nine teeth—Ben spent hour after hour roaring with mirth.

Mabel Barnes enjoyed laughing with Ben. Had she—or any others, for that matter—taken a deeper interest in the old man, they would have realized that Ben was less than happy go lucky. Formerly enslaved in Dublin, Georgia, Ben Ellington was, bit by bit, laugh by laugh, earning money for his family back home. He was also scraping together funds to help his freedom feel real: He wanted to buy a plot of land.

It was a Buffalo reporter who caught Ben off guard for a second,

On stage at the Old Plantation (*left*); "Laughing"
Ben Ellington, seated on right, with fellow performers (*right*).

reminding people why he guffawed so readily: "'Yes, sah, I was a slave (Ha! ha! ha!). Josiah Elander was my master (Ha! ha! ha!). I was on his plantation near Dublin Ga., where I live now (Ha! ha ! ha!). I had a good time when I was good and a bad time when I was bad.'

"Uncle Ben then doubled up like a jackknife and laughed a full minute.

"'I usually gets a dime, gem'n, when I laughs like dat,' said Uncle Ben."[21]

Onto this battlefield then—where white showmen tried to define blackness with savages, smiling slaves, and animals, and where African Americans demanded space to define themselves—strode the figure of James Benjamin Parker. Neither relegated to the Midway nor laughed at, Parker became a champion not only in the Exposition but in the newest, biggest, and most compelling national drama.

And his standing in the public arena lasted about a minute. Almost as soon as he became the most famous African American in the country, it was time to bring him down a peg.

The first efforts to push Parker from his pedestal meant transforming him into a dialect-mumbling man who was full of himself. "If it wan't fo' me," he was alleged to have said, "that mu'derer would of fiahed the rest of them three shots fum his pistol and the President would of bin killed." Parker was also described as an opportunist, walking through the Exposition selling his coat buttons for twenty-five dollars each, and scheming to make more money from shows. It was said he was getting a big head by being feted by city leaders. He was skipping work.

The title of Hero did not adhere easily to a southern African American; it was also extremely inconvenient. Secret Service officers, who credited Parker with fast action immediately after the shooting, soon came to their senses. They insisted they had been perfectly attentive at the Temple of Music. Aside from the one (large) slipup, Secret Service operatives Gallaher and Ireland had been right on top of the assailant. Parker? Jim Parker? The black man? He hadn't really been there at all.

The Seventy-third Coast Artillery, the presidential guard, who were now two for three in public mishaps, also had their doubts. Parker hadn't pulled down the assailant. Nor had a Secret Service agent. It had been their very own Private Frank O'Brien who had leaped into the fray. Or, if not O'Brien, then probably Private Neff, with help from their own Corporal Bertschey and their own Private Brooks. Parker? Parker who?

The *Buffalo Express* summed it up. There was an exclusive club that was getting bigger by the day: "The First Hand on the Assassin Society."

Parker did have his defenders. If the man was missing some coat buttons, asserted one reporter, he gave them away. And he hadn't been feted by city leaders or been dining at the Buffalo Club with rich men. He hadn't even missed an hour at the restaurant. When approached at his job and asked what grand plans he had for the future, he demurred. "What are you going to do?" the reporter asked. "Do nothing," Parker replied. "Ain't I working?"[22]

6.

The Rise and the Fall

I
THE HUB

While Jim Parker fended off a growing number of doubters, other actors in the crisis enjoyed a smooth celebrity. The surgeons who had worked on the president breezed easily into the spotlight. The public and high circles of medicine concurred: The local practitioners had bravely decided to go ahead and operate without waiting for their gunshot specialist, Dr. Roswell Park. Even if they didn't hail from New York or Philadelphia or Boston, they had brought consummate skill to repairing the damage done to the president. They had deftly cleaned his wound, administered anesthesia and painkillers, and done a remarkable job of keeping their precious patient alive. It was an unprecedented achievement. Physicians researching the case claimed that no other ruler who had been shot had actually survived. This was a notable, laudable *first*.

The surgeons rode the crest of public opinion and national gratitude. So did Buffalo itself. The city may have been widely praised for its Exposition, but now, as the headquarters of a national emergency, its capabilities truly shined. From New York City came glowing words. Buffalo, said the *Brooklyn Eagle*, had become "a heart pulsating with sympathy." Its care of the president "was exquisitely well done."

Its residents had "shown themselves patriotic, hospitable, kind-hearted and indomitable."

Buffalo had taken the shooting hard; it could confess this now. The fair had been having its best day ever, the day it (literally) banked on, when McKinley had been shot. Not only did residents resent the fact that their hometown had been the site of such an evil act, but Exposition officials had quietly wondered whether they could survive the blow.

Now, though, the president's expected recovery meant a triumphant new beginning for the city. The Exposition's most intriguing exhibits had simply moved downtown. The Milburn house boasted the latest electric equipment, state-of-the-art telegraph and telephone lines, and Edison's new and sophisticated (and unused) X-ray machine. The house had drawn under its roof many of the country's best medical minds and some of the country's most celebrated journalists, along with their artists and photographers. Famous war correspondents had encamped on Delaware Avenue, as had reporters from top New York City papers. Pulitzer had sent his people. So had the Associated Press.

Buffalo correspondents became dizzy with pride. They stood side by side with some of the country's most illustrious professionals. In fact, one reporter asserted, "a more notable gathering of writers probably never assembled anywhere." The city also seemed to have become the hub of the federal government. The vice president, senators, and cabinet members conducted the important business of the country on its very own streets.

Secretary of War Elihu Root, in town for the crisis, added to the excitement. "Yes," he announced, "Buffalo will be the summer capital." The *Express* echoed him, explaining that cabinet members would be headquartered at the Buffalo Club and the Milburn house would become the temporary White House. Its fame was also assured for the future. "The Milburn home," the paper predicted, "will become historic as the executive mansion as well as the place where the president recovered. In after years the vicinity . . . will be pointed out as a famous neighborhood." Buffalonians, who days earlier had hung their collective heads in discouragement, now beamed.[1]

Pan-American directors concurred with the general sense of pride, but they hastened to remind the public that some of the attention could now be (respectfully) redirected to the Exposition itself. They assured potential guests that news from the Milburn house would be posted on five bulletin boards set up throughout the grounds and urged them not to miss new and original shows. Fresh from the Minnesota State Fair, for instance, an eighteen-foot-tall replica of historic Fort Snelling, made entirely of apples, with crabapple cannons, would soon make its appearance on the grounds. The fort would be almost as dazzling as Minnesota's other marvel, the state capitol building sculpted out of butter.

Fair promoters had more than crabapple guns to help people recover their carefree spirits. Saturday, September 14, was designated as Railroad Day, and, with reduced train fares, it was set to attract more than 150,000 people.

Frank Bostock could hardly be contained. In anticipation of Railroad Day, Jack Maitland, his agent, who had been holding back out of respect for the president, unleashed a new set of gory stories. On September 12, at feeding time, two of Bostock's hybrid lion and tiger cubs fought over a piece of meat, and the male hybrid ripped open the throat of his sister "and drank her blood." Even worse, one of Bostock's most celebrated trainers, Madame Morelli, had been attacked by a jaguar. She had been training him to walk on a tightrope when he "became balky." When she "beat him to make him do his work," the animal jumped on her and tore her arm open. "This is the third time," Maitland explained, "that the same jaguar has attacked her."

Having piqued the interest of fans eager to see bloodstained trainers and vicious animals, Bostock also hoped to produce his delayed and much-touted lion weddings. On Thursday, September 12, he paraded the wedding altar, with lions pacing inside, through Buffalo streets. The next day, he planned to send the altar out again, and, this time, the four brides would be on hand to meet and greet their ferocious witnesses.

Bostock issued regular notices about the wedding. President McKinley had bulletins; Bostock had bulletins, too. The Animal King's last bulletin announced that the weddings would take place beginning at 3:45 on the Exposition grounds, and that the reception would be held in his own arena. He did not name the lucky minister who would officiate at the ceremony but did offer a heartfelt message to the couples: "I know you will never regret being married in a lion's cage."[2]

Railroad Day, featuring the maned and hairy groomsmen, a parachuting human bomb, and an elephant race, would certainly sell masses of tickets and help the fair recover. But officials believed Buffalo could do more. There should be a new sort of President's Day to thank God, thank the doctors, and thank the prayers of the world for the successful fight for the life of William McKinley.

As early as September 10, the public learned that John Milburn and William Buchanan were all for it. It would not be announced, they said, "if there were doubts of the patient's recovery." Doubts or not, news of the planning seeped into the press, and two days later, the director-general was quoted describing the event. The day would be one of "great joy," said Buchanan, a "general thanksgiving over the happy outcome of last Friday's incident." An incident. The attempted assassination of the president had now been downgraded to a short-lived, not terribly meaningful occurrence.

National Jubilee Day, as it would be called, was scheduled for September 21. It would begin with the click of a telegraph key, and this in turn would signal a simultaneous *Hurrah* across the country. Church bells would ring. Factory whistles would steam and toot their joy. Cannons would fire. And the whole United States would resound with happiness.

Within Rainbow City itself, a children's chorus of three to five thousand young singers would add their pure sound to the celebration and, in the Temple of Music (of course), statesmen and orators would speak. One Exposition commissioner proposed (was Jim Parker listening?) that an African American leader be invited to lecture because "one of his race took so prominent a part in the events attending the attempt on the President's life." At this suggestion, other commissioners applauded.

In truth, not all of them clapped. Colonel J. C. Hemphill, representing the upcoming Charleston, South Carolina, exposition, suggested that one of the orators might be a veteran of the Confederate army. That, he submitted, would balance things out.

While the guest list was debated,

Ribbon issued to celebrate President McKinley's anticipated recovery.

what was certain was that Buffalo would be the graceful, hospitable hub of the festivities, the "Mecca toward which the people will be drawn." William Buchanan made it plain that he and others wanted to make Buffalo the place that from then on people visited to express thanks for the president's recovery. As one official explained, it would raise the Pan-American Exposition "from a landmark of gloom to a symbol of happiness."[3]

II

THE TURN

The change occurred on Thursday, September 12. It happened so quickly that those beyond the orbit of the Milburn house might have blinked and missed it. If they were newsmen, maybe they took the afternoon off, perhaps caught up on sleep. If they were city business-men, maybe they read the morning paper, went about their day, and missed the evening news. Or they had a late shift at a factory. Or they went out to the Exposition and came back tired and collapsed from happy fatigue.

And then what they heard made them shudder.

Last thing they knew, the men coming and going from the Mil-burn house were bantering with reporters and saying that the presi-dent was better than ever. He had even asked for a cigar and been served solid food. Nothing pointed to recovery better than a man with a bullet in his back taking in toast, coffee, and chicken broth.

But now? Men were running across the Milburn lawn with their heads down, silent. Automobiles, not horses, were wheeling into view, bringing doctors and nurses.

John Milburn, who hours before had called the president as "fine as silk," began to use a different tone of voice, and made a different sort of announcement. "I do not haul down the flag yet," he said.[4]

On Thursday morning, the twelfth, President McKinley had awoken in a cheerful mood. He had not, however, wanted to eat his toast. By the afternoon, his appetite had gone. He was uncomfortable, restless, and exhausted. His pulse raced—higher than it had been in a while.

The early verdict was positive. The solid food was a mistake, but cathartics surely would do the trick. At 8:30 that evening, however, doctors reported that the president's condition "was not so good. His food has not agreed with him."

Some newsmen wondered whether it was more serious than indigestion. Outside the Milburn house at 10 p.m. on Thursday, they asked Dr. Roswell Park whether he was concerned. "People should not feel alarmed," he replied.

By dawn, Dr. Park was worried too. The cathartic had cleaned out the president's bowels, but strychnine and digitalis and salt solutions had not strengthened his heart or slowed his pulse. Secretary Cortelyou called in Abner McKinley, the present's brother. The medical team asked for more doctors. They sent again for Dr. McBurney, now in Massachusetts, and for the eminent Dr. W. W. Johnston, of Washington, DC. They also summoned Dr. Edward Janeway, who had helped save the writer Rudyard Kipling from pneumonia. Telegrams went out to the cabinet, to the vice president.[5]

Ida McKinley was not told of the downturn in her husband's condition, and Dr. Rixey gave her opiates. Nurses and doctors talked in hushed voices near her room and walked softly.

The weather, on the other hand, could not be muffled. Echoing the alarm, an electrical storm settled over the Niagara frontier, and the sky stuttered with flashes.[6]

Deep in the Adirondack Mountains, Theodore Roosevelt had spent the early part of Friday with his wife and children, and later he gathered a small party of men to climb Mount Marcy. Fifteen miles into the Adirondack wilderness from their base at the Tahawus Club, the hikers had ascended the rocky slopes of the 5,000-foot mountain and were on the way down. They spread out tarps for lunch. It was rainy, but it was the middle of the afternoon, and the men were hungry. They were eating when they heard the noises: cracking twigs, snapping branches, the unmistakable sound of human feet moving fast. A courier broke upon them, a yellow envelope in his hands. Roosevelt guessed, but he tore the paper anyway. "The President's condition has changed for the worse—CORTELYOU." Theodore Roosevelt's hiking trip, his family vacation, and life as he had known it, was over.

Thus began one of the most dramatic changes of command in American history. The rain, begun as mist, became a torrent. After midnight, a buckboard arrived to take Roosevelt off the mountain, and, swaying wildly, its lantern barely hanging onto its flame, it plunged down the pitch. Three times the driver and his passenger stopped at remote stations for fresh horses. Urged to stop and rest, Roosevelt said no. The wagon pulled into the railroad station at North Creek, New York, just as day was breaking, around 5 a.m.[7]

In Buffalo, people had begun to mass near the Milburn house on Thursday night, September 12; and by the next morning, they had to be held back by horses. Part of the reason for the crowds was the news—it was so uncertain.

In the early hours of Friday, September 13, the bulletin delivered to reporters had been hopeful. Then the news came that the president's condition was grave. By 6 a.m., however, his condition had "somewhat improved," and, by early afternoon, the report was even better. He had "more than held his own" and there was "expectation of further improvement."

But no. Four hours after that, at 6:30 p.m. on Friday, came frightening word: the "end is only a question of time."

The watchers passed the time on the street waiting and fearing, sharing stories and premonitions. One among them announced that six weeks earlier he had been to a fireworks show at the Pan-American. As a matter of course, a portrait was lit up in fire—this one of Vice President Roosevelt. It was a good likeness. Then words emerged underneath, in patriotic colors: OUR VICE PRESIDENT, it started to say. But a fuse malfunctioned, and the words *Our* and *Vice* were erased. He wondered whether it had been an omen.

Inside the house, doctors gave their patient all the treatments they had, based on everything they knew: digitalis, adrenaline, strychnine, oxygen, nitroglycerin, camphor, and salt solutions. Clam broth and brandy. Why wouldn't his heart respond? They worried that the first bullet, the one that seemed of so little account, had bruised it. They wondered how strong his heart was to begin with, since he got so little exercise, worked hard, and smoked three to four cigars a day.[8]

In the early evening, when it appeared that McKinley was suffering, doctors gave him morphine and informed his wife that there was no hope. They also determined that she could withstand a last visit, and, in the early evening, before the morphine took hold, the couple had a final, private conversation. Others, too, began to listen carefully to the president. They knew that out of all the words he uttered while conscious, there would be a sentence or two that they would offer to the public as his final comment on life.

Some witnesses thought that what he said about trees was noteworthy. When a nurse rearranged his pillows to block some of the light, he protested. "No, I want to see the trees," he murmured. "They are so beautiful." Others thought his Christian comments were important. He recited the hymn "Nearer, My God, to Thee," and then offered a simple farewell. "Good bye, Good bye all," he said. "It is God's way. His will be done."

Dr. Rixey, on the other hand, reported that the president's very last words were for his wife. "What will become of her?" he wondered.[9]

Across the city, people gathered, warding off the chill and giving way to tears. On Main Street, women and men, well-off and poor, crowded newspaper offices and huddled under electric lights to read bulletins. They were not allowed near Leon Czolgosz. Superintendent William Bull had posted a hundred officers around the city jail, and other men at armories. When asked by a reporter whether he was prepared for riots or a lynching, Bull pounded his hand onto a desk. "We are going to keep this fellow safe, no matter how dastardly his crime may be," he declared. "I advise well-meaning people to keep away from this building . . . and let the law take its course."

At 11 p.m., a rumor circulated—some said it was spread by the press—that the president had died. The word must have reached the city coroner, for he hurried his carriage over to the Milburn house. At 12:30 a.m., he pulled up and asked about the body. He was told to leave at once—the president still breathed.

Inside the residence itself, there were no rumors, just grim truths. Attendants kept the lights on all night, sending rays out into the gloom and making the scene too lively for the hour. In the sick room, Dr. Rixey watched over the ebbing of President McKinley's life as others—mostly relatives, nurses, and orderlies—came and went. The president lost consciousness by 9:30 p.m.; by 11:30, his extremities were cold. His breaths became gasps, the gasps grew uneven, and then they stopped. The people filtered out of the room, the lights went off, and the ever-loyal George Cortelyou went down the stairs. He informed officials in the drawing room and then went out the front door to tell newsmen that the president had died at 2:15 a.m. They in turn passed the word on to the rest of the world.[10]

A few hours later on September 14, Hartford sculptor Edward Pausch telegraphed Cortelyou. He wanted to make a death mask of McKinley. Cortelyou agreed, and at 7:20 a.m., Pausch made his way into the Milburn house. He did not show his eagerness, even though this was the sort of opportunity that might cap his career. He went into the bedroom with buckets of plaster and made a complete mask— not just the face, but the whole massive head. It weighed twenty-five pounds. A man who saw it not long afterward said that the features were "calm and peaceful," and that there were traces of a "benevolent smile about the lips." He also noticed a few of the president's hairs on the right temple of the cast, and, on the right cheek, an eyelash. By November, Pausch's finished work would be on its way to the Smithsonian Institution in Washington. And from then on, anyone wanting to honor William McKinley with a lifelike bust or statue would have Pausch to thank for the model.

On Sunday, September 15, pathologists opened up McKinley's chest and belly. They found, besides a heart that was weakened by age, an insidious bullet track. The missile had entered and exited the stomach, nicked a kidney, but, worst of all, damaged the pancreas, spilling its poisonous enzymes. The track had become a gangrenous tunnel. A century later, surgeons employing antibiotics, imaging, and intravenous fluids could have saved a man injured like this, but in 1901 it was an almost impossible task.

The physicians spent four hours looking for the bullet that had killed the president but, try as they might, and digging as deep as they could, they could not locate it. Ultimately, they were asked to stop searching. The family did not want "to injure the corpse any longer."[11]

Theodore Roosevelt arrived in Buffalo at a little after 1:30 on Saturday afternoon, September 14. He alighted from the train, brushed aside some young boys who had broken through the guards, quelled cheers, and shook hands. To a reporter's question, he simply replied, "There is nothing to say at all." He was then walled in by police and mounted soldiers, and the entourage moved north. He stopped at Ansley Wilcox's home briefly, then moved up Delaware Avenue to pay his respects to the McKinley family.

By midafternoon, Roosevelt was back at the Wilcox house, and at

President Theodore Roosevelt, newly
sworn in, talks with reporters.

3:30 p.m., in borrowed frock coat and hat, he stood in the bay window of the house's library and took the oath of office. When Secretary of War Elihu Root, on behalf of the Cabinet, formally asked Roosevelt to take the oath, Root got out only three words before he broke down. For two minutes he cried. Others cried too. And then Roosevelt signed a piece of parchment and the deed was done.

Farther north, at the Pan-American Exposition, the fairgrounds had become a graveyard. The big, empty buildings and the silent, dark towers were, said one man who walked in, "monuments to a dead president." Exhibits were locked, gates were closed, and the Midway for once was nearly soundless. Like somber apparitions, a few employees flitted in and out of the gates, did some work, and scattered.

Dusk signaled the time for the Illumination, that moment when the Exposition renewed itself with electric light and urged guests to stay late and revel as though the day were young. This night, though, there was no returning. No one made a mistake and tried to pull a switch or push a lever. The dark descended on Rainbow City and stayed.[12]

7.

Aftershock

I
RITES AND PASSAGES

The Exposition remained closed from Saturday, September 14, until Monday, September 16. A man who walked through the grounds said the only sound he could hear was hammering, as one exhibitor after the next nailed up portraits of the dead president.

On Sunday, the city that had ten days earlier been dressed in starred-and-striped bunting stood sheathed in black crepe. At midday, ebony-colored horses pulled William McKinley's coffin four miles from the Milburn house to Buffalo City Hall. Ahead of the hearse, a band set a melancholy rhythm to Chopin's "Funeral March," while, behind it, Civil War veterans, gray and unsteady, marched to honor their comrade. As the procession passed Trinity Church, choirboys emerged to sing "Nearer, My God, to Thee." McKinley had uttered the hymn, or tried to, just before he died.

The cortege arrived at City Hall just as clouds driven by a southwest wind off the lake unleashed a downpour. Sailors from the battleship *Indiana* and soldiers from the Seventy-third Company, Coast Artillery, who had been so much a part of the president's time in Buffalo, carried the casket into the building, set it down on a catafalque, and opened it. William McKinley lay on a silk pillow, his face under glass. His left hand, which had slipped off his chest during the trip

President William McKinley lies in state, Buffalo City Hall.

downtown, was repositioned. One observer thought that his face was somewhat sallow, but that all in all he was only a little "emaciated."

From early afternoon until late at night, mourners, smelling of wet wool and dripping water, shuffled in two by two and passed the bier. The new president and the cabinet paid their respects, along with politicians and businessmen. The guiding lights of the Exposition came too, as did commissioners from Latin America. And the people came, many from Erie County but others from Cleveland and Pittsburgh and New York. And the Midway came—Bedouins in "Sahara costume," and more than a hundred members of the Indian Congress. A note, said to be from Geronimo, sat next to a wreath of purple asters. "The rainbow of hope is out of the sky," it read. Geronimo himself walked through the rotunda with other Apache, as did Lakota, including Red Cloud and Painted Horse. Shot-in-the-Eye, who had been at Little Big Horn and had seen the end of George Armstrong Custer, joined them.

And Jim Parker came, wearing a wide black armband. People murmured as they recognized the tall man and watched him bend low over the coffin to have a long last look. Parker had not been able to shake the hand of the president nine days earlier, but he was at least able to honor him now.[1]

William McKinley's body left Buffalo as the man had arrived, by train. At 8:30 Monday morning, September 16, five funeral cars, including a glassed-in observation car carrying the casket, rolled out of town toward the nation's capital. The train moved southeast, threading its way from New York into Pennsylvania, where it picked up the Susquehanna, and headed into Maryland. Miles and miles of mourners waited along the railroad bed: laborers and housewives, schoolchildren and office workers. Farmers, silhouetted and still, stood in their fields like cairns against the sullen sky. People had strewn flowers on the track—roses and asters and violets—and, as the cortege rolled past, they sang hymns and rang church bells. It wasn't enough to see or try to photograph the train; a few of the crowd, mostly young boys, put coins on the rails, so that they could forever feel the weight that flattened them.

Ida McKinley traveled in the second coach, alternately resting and staring out the window. Once in a while, especially when she caught the sound of a hymn, she would begin to weep. The new president and his advisers rode along as well, and, in a tribute to Buffalo, four city leaders, including John Milburn and Conrad Diehl, had been asked to join the group, and were part of the funeral services from start to finish.

After ceremonies in Washington, where McKinley lay in state at the Capitol and tens more thousands wound through the streets to bid him good-bye, the entourage journeyed back to Ohio. This time, the train traveled after dark through coal country, past mourners warmed by bonfires, and miners, their headlamps lit, standing bowed. Near Pittsburgh, people on a bridge tipped three wagons' worth of flowers onto the tops of the funeral cars. By the Allegheny River, men and boys hung onto bridge girders to see the train pass. Some people in Pittsburgh, more angry than grief-stricken, dangled an effigy of Czolgosz from a factory crane. The knives they plunged into the body glinted in the electric light.[2]

The train arrived in Canton on Wednesday at noon, and the president's body was carried into the courthouse for a final viewing. Local officials insisted that the casket be opened, and then regretted it. When a chandelier's light was directed onto the president's face, the shock was audible. It was, said one reporter, "pitifully unfortunate, as the discoloration had increased." Witnesses emerged from the scene, sobbing. They had, perhaps, expected to see a man safely in repose, ready for eternity. Instead, they looked at the decay of death.

Ida McKinley wanted to see that face, too, to have a last good-bye, but no one was willing to open the coffin for her. She did persuade officials to let her husband's body lie at home for a last night, and there she remained sequestered, unwilling or unable to attend the church funeral and the burial. When the president's coffin was carried out of the house, an observer below thought he saw her. Someone pulled back a curtain on the second floor, and a frail, mothlike face appeared, looked down, then faded away.

At 4 p.m., the twenty-fifth president of the United States, followed by eleven flower-filled cars, was carried to Canton's Westlawn Cemetery and placed in a family vault. As they had at each event in this long ritual, the Buffalo representatives stood in line with President Roosevelt and members of the cabinet as the casket moved past. For days, these men had been honored members of the funeral party. Roosevelt had gathered them together on the train and thanked them for "the bearing of the people of Buffalo after the terrible tragedy." What they had offered and done, he said, "was appreciated more than words could make plain."[3]

II
FAULT LINES

The funeral trip had been hard, but now the Buffalo men faced a new challenge: They had to go home. They had to pick up the pieces of a city in shock, steer the public away from its grim fixation with the president's death, and rescue the Exposition.

They also had to deal with the city's recent grandstanding: its pride, its sense of real importance when McKinley seemed to be recovering. They had to contend with the fact that some commentators blamed Buffalo for McKinley's death. These critics said the police department hadn't done enough to protect the president and suggested the city's surgeons had been bumblers. The physicians had made a "startling error of diagnosis" and been overconfident. "What excuse must be offered to the public for the utter inability to find the bullet even in the dead body?" asked New York surgeon George Shrady in the *Medical Record*.

Others accused the Buffalo doctors of being unprofessional. They had been optimistic when they should have been circumspect and cautious, and they had stooped to undignified squabbling. They had disagreed over the path of the bullet, over the meaning of the president's high pulse, over whether or not a poison had caused the gangrene. They had "talked too much." A Manhattan doctor thought the finger-pointing could have waited at least until after the president's body had made it to its resting place.

The surgeons fought back. There had been no disagreements, no unprofessional behavior. They condemned such "scandalmongering," particularly from the New York yellow press. Another medical journal entered the fray and leapt to Buffalo's defense. How could anyone insinuate that the tireless physicians of Buffalo had not done everything they could? They had, after all, kept the president alive for a week. These innuendos had "stung the whole medical fraternity" and turned McKinley's death into a "popular spectacle."[4]

Perhaps worse than the reproach, though, was the pity.

The *New York Times* reported that Buffalo was utterly shattered. Even before the Exposition, it noted, "there was no city in the country where friendship and devotion for [McKinley] were more marked." And after it had poured its heart and soul into the Exposition, the city had then experienced ten days of worry and heartache. Now it had sunk even deeper, into depression "beyond expression."

City officials and Exposition directors could not deny that the assassination hovered over the fair. The gates opened again on September 16, and more than sixty thousand visitors filed into the grounds. They went dutifully to cattle shows and watched Short Horns and Dutch Belted and Holsteins win money and silver cups. They stood by a wireless telegraphy demonstration and stayed for the fireworks at the Park Lake. And they applauded when applause was called for. But their enthusiasm seemed forced, as though they were trying to exorcise the sadness. Everybody knew—the newspapers were full of it—that Czolgosz was close by, still eating and sleeping well, and that maddened them.

Showmen felt the gloom too. A week after the president's death, the *Buffalo Commercial* reported that performers at the Indian Congress had sensed bad spirits flickering through the fairgrounds. Chief Seven Rabbits, medicine man of the Congress, condemned the spirits and asked his followers to pray—for snow. He believed "that snow would be the only thing to drive bad spirits away in this climate."

Buffalo was good at snow, to be sure, but in mid-September? Even Buffalo couldn't answer that prayer. Meanwhile, Director-General Buchanan was receiving bundles of telegrams. People were wondering whether the Exposition was closed for good.

Determined not to bury the fair along with the president, John

Milburn begged people not to give up on the show. He reminded the public that the Exposition was a tribute to McKinley and his policies. Others echoed him, suggesting that McKinley was martyred for the fair, that he had given his life supporting the show, and that the Pan-American was now a McKinley shrine.

In fact, a shrine of sorts was in the making. A group of Ohio visitors, surveying the assassination site in the Temple of Music, made a startling find. Almost in the very place where McKinley had been shot, in the grain of the wood floor, they found "a splendid likeness" of the dead leader. If visitors looked at the image in the right light (afternoon was best), they could see McKinley's "high forehead and Napoleonic nose," not to mention his "heavy overhanging eyebrows." The face in the wood attracted scores of sightseers.[5]

On the Midway, the Animal King rallied to the cause. True, his lion-cage weddings had been derailed by the death of the president. Two of the fearless couples had not been able to reschedule and were married tamely in the Sunday School Room of the Central Presbyterian Church.

But Bostock announced more weddings, and added other shows. In a burst of optimism, he repainted the exterior of his exhibit hall in the colors of the rainbow. And he bought more animals: a new hyena; a Tasmanian devil; two polar bears; a red river hog from Egypt; a herd of trained zebras; and a troupe of baby elephants who could play on a seesaw, walk on their hind legs, and roll barrels. The showman also printed twenty-five thousand postcards picturing himself, Madame Morelli, and Captain Bonavita sitting with a perfect pyramid of twenty-seven lions. "The Climax of Animal Subjugation," the caption read. A one-cent stamp sent the card across the country.

There was very little that ruffled the confidence of the Animal King—including, it seems, the man who paid for a ticket and entered his lion show on September 12. We do not know whether a report triggered the stranger's visit, or whether it was just routine scrutiny, but the Humane Society agent took a seat in the crowd at "The Training of Young Lions," one of the most popular wild-animal acts. He watched for a bit, eyeing the animals carefully. What he saw disturbed him. The trainer used a sharp iron prong to deliver discipline, and at least one young lion was bleeding. The agent brought formal charges of cruelty against the show. Bostock was called into court to testify, and, genial as ever, promised to promptly remove the act in question.[6]

III
THE VERDICT

On September 16, the American public, not to mention Buffalo residents, received some sense of justice done when the grand jury of the Erie County Court indicted Leon Czolgosz for murder in the first degree. Czolgosz appeared at the arraignment in a rumpled linen shirt and a new beard. His hair, no longer pomaded, looked lighter and more unkempt. He came into the courtroom, looked at the ground, and stood mute.

Thomas Penney, the district attorney, got down to business quickly.

"Czolgosz, have you got a lawyer?" he asked.

There was no response.

"Have you got a lawyer?" he asked again.

Silence.

"I say, Czolgosz, have you got a lawyer?"

Nothing.

He was impatient now. "Czolgosz, you have been indicted for murder in the first degree. Do you want counsel to defend you?"

Nothing.

Penney moved closer to the prisoner, almost face-to-face: "Czolgosz, look at me and answer."

Still nothing.

One reporter who watched the procedure said that the prisoner stared emptily, as if he were "insane." Others interpreted the vacant looks as subterfuge. He was *pretending* to be insane. Or perhaps he was acting out an anarchist script. Criminal anarchists had been known to keep silent until all hope of avoiding the death penalty was past, then jump up and shout their vile beliefs.

The ensuing murder trial brought the city of Buffalo high praise. By most accounts, it was a model of speed and efficiency and demonstrated America at its most civilized. If it had been performed as an exhibit at the Pan-American, John Milburn and company would have been delighted.

Beginning on September 23, the quickly chosen jury of twelve men considered a long parade of experts and witnesses for the prosecution. Judge Loran Lewis, on the other hand, told the jury that his assigned job as Czolgosz's defense attorney was "exceedingly unpleasant." He explained that there was no denying that the defendant had shot the president. The only question was whether or not he was sane. If he were sane, then he was guilty and must be punished. If he were not, then he would be acquitted and "confined in a lunatic asylum." The problem, he went on, was that the defendant had, for the most part, refused to speak to his court-appointed lawyers.

Lewis did make a small attempt at an insanity defense. All people, he asserted, wanted to avoid death. And yet this man had done a deed that he knew was certain to cause his death. "Could a man with a sane mind perform such an act?" he asked. He suggested there might be benefits to finding the defendant insane. If McKinley had been shot by a madman, the nation would breathe a collective sigh of relief: It wouldn't have to worry about "a conspiracy of evil disposed men" who might want to hunt down other rulers, even another president of the United States.

Lewis did not enlist the help of any witnesses to make his case; nor did he call on any one of the five psychiatric experts who had met with Czolgosz. Dr. Carlos MacDonald of New York City's Bellevue Hospital might have helped the defense, admitting as he did that the psychiatric history of Czolgosz had been rushed, and that physicians had been unable to interview the accused man's relatives.

Instead, the defense attorney, a veteran judge with the New York Supreme Court, directed the jury's pity to the fallen president. "He was a tender and devoted husband," he said, "a man of finest character and his death is the saddest blow I have ever known." As he spoke, his voice shook, and he had trouble withholding his tears. Meanwhile, Czolgosz sat stonelike, his head tilted. Flies landed on his forehead and he made no effort to brush them away.

Prosecutors maintained that Czolgosz's sanity was demonstrated by the fact that the defense could not prove—did not even try to prove— otherwise. They were thus spared the pain of discussing what had moved Czolgosz to anarchism. They did not have to listen to the assassin talk about the division of wealth in the country or about the condition of poor people. Ideas like these might have pointed to the man's sanity, to be sure, but—this was the tricky part—they might have elicited an emotion that no one, defense or prosecution, wanted to see in the eyes of the jury, and that was sympathy.

It was over in a flash. The jury left to consider the evidence at 3:50 p.m. on September 24 and returned just over half an hour later to deliver the verdict: guilty of murder in the first degree. The sessions of the trial had lasted less than nine hours—including an hour and a half to select a jury. A few days later, Czolgosz was sentenced to die by the electric chair at the end of October. After hearing his sentence, the prisoner returned to jail, and, as usual, ate a hearty meal. He then went to bed and slept soundly.[7]

The proceedings of the trial used up nearly every word of admiration in the lexicon. Only a few commentators, including Emma Goldman and some psychiatrists, described it as farcical.

IV
THE EXCISION

While Buffalonians were pleased with the murder verdict, many local African Americans read the trial transcript with suspicion. The prosecution had called many witnesses to the stand, including soldiers, Secret Service men, and city police. The lawyers had not, however, been interested in hearing from Jim Parker. Sometime between the shooting in the Temple of Music and the indictment and trial, Parker had been erased from the assassination's record of events. Secret Service agent George Foster, involved in the takedown of Czolgosz, when asked specifically about Parker, testified that he "never saw no colored man in the whole fracas."

The act of ignoring Parker did not go unnoticed. Across the country, some newspapers picked up on the slight and quickly said, "We told you so." At the same time, white Republican papers rose to defend Parker, rallying evidence. If he hadn't actually intervened, why had every paper labeled him a hero right after the event? Why had Agent James Ireland, in his report to the chief of the Secret Service, named Parker as a man who had grabbed the assassin? Why had Parker originally been named as one of the victims of Czolgosz's assault? Other witnesses testified to his presence at the Temple of Music, including two white southerners who had been at the fair.

On the night of September 27, a number of Buffalo's African Americans met at the Vine Street African Methodist Episcopal Church to weigh the conflicting reports. While a committee convened in a separate room, men rose to speak. One, named Shaw, had worked with Parker at the Plaza Restaurant and argued that the effort to "discredit" Parker was the result of a plot. "Should we fail to emphatically resent it," he contended, "I claim we are a disgrace to our race."

As Shaw was speaking, Parker himself arrived at the door of the church, entered, and moved down the aisle. The audience swooned. Women fluttered handkerchiefs and men whooped, tossed their hats, and boomed three cheers.

The committee reentered the room, discussed the evidence, and read out two resolutions. One lamented McKinley's assassination. The other, directed at Parker, equivocated: "It is the sense of the colored citizens of Buffalo, N.Y., in mass meeting assembled, that they very much regret the clash of statement in respect to the reported act of heroism on the part of James B. Parker." The committee members regretted it so much, apparently, that they could not offer an opinion, and definitive word would have to wait. It would be up to historians, they said, "to award honor to whom honor is due."

But while the committee held back, others who were there did not. In giving three cheers for Parker, in urging him to speak to them, and by surrounding him on the street after he left the church, they told him they believed him.

Soon afterward, Parker did what he could on his own to set the record straight. He went beyond Buffalo, lecturing about the assassination throughout the East, and he visited President Roosevelt at the White House. While in Washington, he explained to a gathering at the Metropolitan AME church how he had come to the defense of the president. He told them how District Attorney Thomas Penney, for the prosecution, had interviewed him early on but then, somehow, had lost interest in his opinion. "I don't say that this was done with any intent to defraud me," he explained to the group, "but it looks mighty funny, that's all. Because I was a waiter, Mr. Penney thought I had no sense. I don't know why I wasn't summoned to the trial." A woman in the audience had a ready answer for him. "Cause you'se black; that's de reason," she said.[8]

V
GERONIMO

Back in Rainbow City, temperatures were dropping, and on the Midway, performers found themselves unready for the cold. In Darkest Africa, the frigid weather forced villagers into borrowed clothes, and visitors found the sight ridiculous. The Africans also tried to warm themselves around a gas stove, and when one used a match and caused an explosion, singeing himself, it was another occasion for laughs. At the Old Plantation, the cold meant more entertainment, as "the darkies dance and sing with added vim."

From the Indian Congress, Geronimo wasn't amused, and he made a plea for seven hundred overcoats. At the Filipino Village, the cold weather aggravated the tuberculosis of a young woman, and she died.[9]

While occasionally turning their backs on performers' needs, concession managers continued to use them to make money. And nothing made money more spectacularly than the show planned for the end of September by the Indian Congress. On September 26 and 28, up to twenty thousand guests took their seats to watch an Indian dog feast.

Frederick Cummins, manager of the Indian Congress, orchestrated the killing and eating of the dogs, but most people credited Geronimo with the inspiration. Throughout the summer and early

Apache leader Geronimo and Wenona, the "Sioux" sharpshooter, center,
pose with visitors and other members of the Indian Congress.

fall, the Apache leader had become the most visible Native performer
at the fair. Once known by the US Army as the "human tiger," the
warrior chief, like many other non-European men on the Midway,
was now pictured as entirely tamed, even feminine. He was said to
be vain about his appearance, fussy about his hair, and expert at bead
designs. Word had it that he had taken up a musical instrument and
had begun offering cooking classes.

Geronimo seemed so removed from his former life as an enemy
warrior that he not only played an Indian in sham battles; he also
offered white guests a chance to play Indian. Without taking off
their top hats or changing out of their shirtwaists, and with enough
money—usually seventy dollars—visitors could be inducted into a
tribe, usually Apache, Oglala Sioux or Arapaho. They received funny,
pretend-Indian names. Railroad executive Charles Clark of the Big

Four Railroad became "Chief Likes His Eggs"; Captain Hobson of the Queen & Crescent Railway became "Chief Blows Them Up."

Geronimo had performed agreeably in all of these ceremonies—he was, after all, still a prisoner of war—but, especially as the season wound down, his irritation at curious crowds seemed to grow. He began to charge more and more money for photographs, and, one day in September, he lost patience altogether. When a spectator tried to follow him for a photograph, Geronimo "buckled into him, tipped him over, [and] knocked his camera to one side."

Geronimo wasn't alone. Other Plains Natives may not have acted out in anger, but they turned jokes on their audiences. The Indian Congress regularly featured a line of performers giving "speeches" in Native languages. These speeches tended to be a commentary on visitors, criticizing their clothing, their appearance, and their mannerisms. Behind them, other Native men listened and laughed.

Now, Geronimo and seven hundred other Native people had

At the Indian Congress, fairgoers peer under the flap of a tepee.

something different to offer Exposition crowds. Perhaps they were unapologetically or defiantly performing a meaningful ritual. Or perhaps they were performing a "savage" act to bring in revenue. Likely they were doing both.[10]

Dog feasts were nothing new for some Native peoples. Northern groups especially, like the Lakota, Cheyenne, and the Bannock of Idaho, held dog feasts on special occasions, as did a few southern tribes like the Osage, and the Sac and Fox. Occasionally, tribes fattened their own animals for the sacrifice, while others collected stray dogs from beyond their encampment. Women often did not partake.

Not surprisingly, members of the white press had weighed in on these events with disgusted sighs. What had the missionaries, the teachers, the churches accomplished if these people were still eating "pets"? At the same time, dog feasts reinforced reporters' own sophistication. And, they had to admit, they were fascinated.

The Buffalo dogs were taken quietly at first. Managers of the Indian Congress bought strays from dog pounds and dogcatchers, and they emptied neighborhoods, streets, and parks. By September 22, three hundred dogs, including poodles, terriers, and spaniels, had been corralled. The Congress, though, wanted four hundred more.

In Buffalo and nearby towns, city residents, eager for the money or the attention, offered up their own animals. In Tonawanda, a woman named Johnson offered her five adult dogs and seven pups. She needed the two dollars. And a Mrs. Foster of Elmira sent a telegram to Henry P. Burgard, president of the Indian Congress, stating that she had a pug named Lillian Langtry that she would part with "if she was certain an Indian would eat it."

Newspaper accounts like these prompted Buffalo's reform-minded residents to protest. In fact, the Humane Society would see its failure to stop the dog feast as one of the year's great disappointments. When officers of the society called on the management of the Indian Congress to halt the event, they were shown permission records issued by the United States Government. There was nothing the society could do.

The animal protectors also admitted they were conflicted. Humane Society officer Matilda Karnes explained that respecting Indian visitors meant acknowledging that the "religious festival must be considered legitimate." It was the "degradation which was forced upon us," she said, that was humiliating. The society had labored long to exert a positive influence over young people and this "barbarous" event was destroying its good work. The group knew the counterargument: Killing dogs was not a lot different from slaughtering other animals for food. But, Karnes said, that was not the point. What the society condemned was not the killing, but the spectacle of killing.

At 5 p.m. on September 26, before an audience of at least ten thousand, the sacrifice began. Geronimo killed a dog with a bow and arrow, and then sharpshooter Wenona took her turn with a rifle. Newspapers carried every detail, from the killing of the animals to the eating of them. They reported that Geronimo ate two dogs. They tasted, he said, "like fried frog legs."[11]

Frank Bostock did not comment publicly on the commercial success of the dog feast, but he was just yards away from the event. He made his living showing off his mastery of animals. Was there money to be made in the death of one?

Not unrelated to that question was the ongoing arrival of animals at his compound: a new llama, a vicuña, a razorback hog, and a black snake, sixteen feet long. There never seemed to be an issue with overcrowding in the arena. And when it came time to sell animal skins at the end of the Exposition, Bostock had plenty to offer.

Bostock's Doll Lady, meanwhile, worked through the fall without complaint. Tony, on the other hand, had grown tired of their secretive routine. One night, he crawled through the window of Chiquita's dressing room and steeled his nerve. He told her he loved her, and asked how she felt about him. Alice told him she loved him, too. "Do you love me well enough to marry me?" asked Tony. She said she did.

Tony wanted to get married then and there, but Chiquita stopped him. She was afraid of Bostock. She was, however, willing to make a date to elope: Friday night, November first. It was the night before the Exposition closed, so it seemed safe enough. Tony enlisted his brother Eddie to help with the engagement by purchasing a ring, and shortly afterward he hoisted himself over the roof of the Ostrich Farm and sneaked into his sweetheart's quarters to present it.[12]

VI
THE DARK MARK

On September 25, a Buffalo newsman estimated that of the fifty-eight thousand at the fair the day before, less than two per cent were local visitors. He accused city residents of forsaking their fair, noting that while the rest of the country was being asked to go to the Exposition out of patriotic duty, city residents had been let off the hook. It was time they did their part. "If Buffalo people turned out as generously as Chicago people did at the World's Fair at this time of the season, the attendance would be nearer to 100,000."

Residents rallied. September 28, Railroad Day, which had been postponed when McKinley died, brought in nearly 120,000 people. Mabel Barnes showed up, of course. Back at school now, and forced to squeeze her visits into weekends, she had been to the fair only twice since McKinley's death. Now, arriving alone, she was on a mission to see every last bit of the Exposition. She had state and country buildings on her list, art and sculpture to see, and she wanted more music, more parades, more Midway. Years later, when Rainbow City had been reduced to splinters and dust, and Mabel sat assembling all her notes, perhaps she took pride in the fact that she, above all others, had been flawlessly dedicated.[13]

While the loyal schoolteacher made her way speedily through exhibit buildings on Railroad Day, other visitors watched a man named Leo Stevens, "the human bomb," ascend into the sky in a big round ball, explode out of it, and parachute back to the ground. A good many of them also took the opportunity—finally—to witness a young couple marry in the company of lions.

At 4:45 in the afternoon, the betrothed couple, Caro Clancy and William McAlpin, waited while Bostock's lion trainers marched over the Triumphal Bridge, followed by the altar. The cage stopped in the middle of the Esplanade.

Miss Clancy, who wore a white gown, a feathered hat, and carried orange blossoms, and whose cheeks betrayed a tinge of excitement, climbed into the cage and stood between her future husband and the lions. The minister positioned himself just inside the door and began

The long-awaited lion wedding.

to read the service. The trainer waved his whip up and down. With great efficiency, the minister led a prayer and produced a certificate. Then he edged to the door. The lions roared. "Hurry up, hurry up," a witness shouted. The groom and the minister moved out of the cage first, and the bride followed them. "She was," said an observer, "first in and last out."[14]

It is hard to know whether never-ending comparisons with Chicago's world's fair bothered or pleased Buffalonians. What is certain, however, is that October 7, Illinois Day, generated more than the usual fuss. Illinois governor Richard Yates, Senators William Mason and James Templeton, the mayor of Chicago, and Chicago aldermen would be honored with parades, a military review, banquets, luncheons, and a tour of Niagara Falls. The supporters of Rainbow City would show the former denizens of the White City how proud they were of their own production.

The festivities got off to a shaky start. For unclear reasons, the Illinois regiment serving as the governor's escort traveled from Chicago to Buffalo on October fifth with almost nothing to eat, so the soldiers arrived hungry at Buffalo's Exchange Street Station. A two-mile march to the center of the city and a train ride to the Exposition, again without food, did nothing for their frames of mind. Aboard the train, they sang loud songs about chicken, beef, and hot dogs. "Hallelujah give us a sandwich to revive us again," they crooned. They also pretended to be conductors. "The next station at which this train stops is Dinner Avenue," shouted one soldier. Another claimed, at the Auditorium stop, that he could eat the building's bricks. Thirty hours after leaving Chicago, they arrived at the grounds and made a beeline for Bailey's kitchen.

The mayor of Chicago couldn't attend the event. But city aldermen made their way east in large numbers, and Governor Yates arrived in Buffalo early on Sunday, October 6. After attending church, the governor's party of twenty-seven decided to tour the Niagara River along the Canadian side. They were on their way back by rail, heading toward the falls, when, a little after 4 p.m., their car lurched, smoked, and stopped dead. There were, someone recalled, "a few screams."

The governor and his fellow passengers, all uninjured, held a conference and decided, with no help imminent and no way to call for help, that the solution was to get out onto the tracks and walk the rest of the way to the falls. Two to three miles, they were told. The young, athletic governor was especially undaunted. He had, as a college student, walked seven miles before breakfast, so this was nothing.

But it wasn't a couple of miles. It was six. As the group trudged along the rails, the sun went down, and then set. The men and women had nothing to eat. The roadbed, which skirted cliffs and moved through narrow trestles, was rough. The party straggled into Niagara Falls, took a working railroad car to Buffalo, and, around 8 p.m., went straight to their hotel.

There was more awkward walking the following day, because fair officials forgot to tell the Illinois visitors how long it might take to get from their hotel to the Exposition. But good humor prevailed, and the ceremonies in the Temple of Music moved forward without mishap. The visiting speakers thanked Buffalo, congratulated the city on its Exposition, and only discreetly alluded to the Columbian Exposition.

But they might as well have waved a White City banner alongside their regimental colors, for the theme that rose above all others was Buffalo's—it seemed to belong to Buffalo—sad assassination. Senator Mason, after talking about the virtues and beauty of Illinois, offered condolences. Americans knew that Buffalo had been kind and attentive to the president, he said. They also remembered "that the blow at the nation was struck at Buffalo."

Governor Yates graciously offered thanks to his audience and explained that Illinoisans knew how they felt. Lincoln, America's greatest hero, and an Illinois native son, had fallen to a madman's bullet. They felt for Buffalo. Edwin Munger, chairman of the citizens' committee of Chicago, added that his group had come to Rainbow City "with saddened hearts and drooping spirits." Now here they stood, he said, "in the almost visible presence of the most awful and most senseless crime of the century."[15]

For all of Rainbow City's artful colors and lights, its grand exhibit halls, its celebration of the Western Hemisphere, and its Midway, what stood above them for these Chicagoans was Buffalo's tragedy.

8.

Freefall

I
SEESAW

As October progressed, mood swings at the Exposition accelerated. There were days that seemed to bring Buffalo residents unadulterated joy, such as New York State Day on October 9, when nearly 130,000 visitors broke the attendance record. The following week, though, a *Buffalo Times* headline announced: "Buchanan Admits Loss," and a column declared that original Buffalo stockholders "will in all probability realize nothing." The next morning, Fidelity Trust Company, which held the Exposition bonds, opened its windows to nervous customers. They came quietly at first to withdraw their money, but, as rumors of a failure spread, they formed lines that snaked into the streets. Bank officers tried to calm them, pointing to tellers who sat behind windows with towers of cash at the ready.

Judge Loran Lewis, a director of the bank and the man who three weeks earlier had defended Leon Czolgosz, got up on a chair to speak to those in the bank's lobby. "You ought in the interests of Buffalo," he said, "go away and do all you can to allay this absurd excitement. You should not have allowed yourselves to get into this foolish flurry." Some people listened and went away. Others made deposits, even big deposits. Some customers, though, more concerned about their money than the image of the city, went up to the tellers, took away cash, and went home.[1]

At Auburn State Prison, 150 miles to the east, the man responsible for much of the misery in Buffalo continued to eat big meals and rest soundly. He also sat, paced, and, infrequently, talked with guards. He asked for nothing to read, nothing to write with, and wanted no consolation. Because he received no news, he did not know that anarchists in London had labeled him "Saint Czolgosz" and that in New York an anarchist had delivered a public rant denouncing McKinley's policies and linking the assassination to "the oppression of workmen by capital."

The prison warden, Warren Mead, let it be known that Czolgosz remained calm about his upcoming death. He did, however, reveal some details about the plans for the upcoming electrocution, including the fact that he had received more than a thousand requests to witness the event.[2]

For Exposition officials, October 19—Buffalo Day—signaled a big, final beam of hope. The press reminded people that Chicago Day had brought seven hundred thousand people to the White City, and they urged everyone to pitch in—close a business or shop, ask for the day off, head for the grounds—make this a "landmark" in Buffalo history.

The event got off to a good start, and more than ninety thousand people had entered the grounds by 1 p.m. At the football game, between the Carlisle Indian School and Cornell University, people crowding the stadium gates fell and were stepped on, requiring Exposition police to use their clubs to restore order.

The Exhibition halls and the Midway were mobbed. There were

so many crowds outside Bostock's show that the Animal King did something he hadn't done all season: He called off advertising. He sent spielers home and asked the police to come. He was beside himself with delight. It was, said one reporter, "the Midway showman's realization of earthly bliss." And it allowed him a grand boast: "The attendance at the animal show yesterday breaks all records for any Midway concession at any Exposition."

The final number, 162,652, meant that Buffalo Day had surpassed expectations. And yet . . . and still . . . Pan-American directors knew they needed more people to spend more money. They added two days to the season to add revenue and would close on November second. For his part, John Milburn let on that he was open to new ideas for Farewell Day. "Just at present I am cudgeling my brain," he said, for an event to "arouse popular interest." Concessionaires also searched for a different act for Farewell Day. "They think," said one newsman, "that by putting on some sensational features of an entirely new stripe . . . they can bring in trainload after trainload of people."[3]

They wanted something wondrous that would draw thousands. And above all, something altogether new. They had run through all the "Days" in their repertoire.

But could they ever, in their wildest dreams, have conjured up the stout, dowdy, big-hatted woman who hoped to be the answer to their prayers?

II
THE NOMAD

In early October, it seemed as though Annie Taylor would not even make it to the Niagara frontier. West Bay City Cooperage had barely begun to work on her order. And the barrel, such as it was, didn't seem

Annie Taylor poses with her barrel and a cat.

fit to do the job. A local newspaperman who called on the cooperage to survey the construction thought it wouldn't hold much oxygen. Furthermore, he wondered whether Taylor would fit. She would have to "snuggle herself" into a space four and a half feet high and twenty-four inches wide. He did admit the barrel looked sturdy. Made of Kentucky oak, its staves were more than an inch thick and its iron hoops were bolted through the wood. But he remained unconvinced.

Yet less than a week later, the barrel, with an oiled and waterproofed cover, was done. Tussie Russell drummed up publicity by putting it on display in a Bay City storefront on October 8 and announced that he would leave Michigan the next day to make arrangements. Four days after that, on the evening of October 12, Annie Taylor stepped aboard a train and headed east on the Pere Marquette Railroad. Calm and unwavering, she bore no sign that she teetered on the edge of a self-made disaster. Or that the gruesome death of Maud Willard, just a month earlier, had dented her resolve.[4]

The route that had brought Annie Taylor to this moment was neither straight nor smooth, but if we believe her, she was one of the most intrepid adventurers of her time. IF we believe her, that is. Later, people would discover that she lied about one of the basic facts of her life.

We do know Taylor had been born as Anna Edson in Auburn, New York. The daughter of a well-to-do miller, she had grown up in a large family. She said she enjoyed the roughhouse company of her brothers and, sitting in the woods by herself, dreamed of a future befitting a boy: pursuing adventures and seeing distant places. She read Sir Walter Scott's work, as well as stories about pirates, Indians, Ancient Rome, and the Australian outback. "My brain," she said, "was teeming with Romance."

But real life intruded: her parents' death before she was fourteen, and worries about money. Her father's estate would not be divided until a younger brother came of age. She accepted help to attend a teachers' training school in Albany and married a man named David Taylor in her late teens. She bore a son who died as an infant, and, after two years of marriage, was widowed.

Soon afterward, around the middle of the century, Annie Taylor traveled to San Antonio, Texas, to see a school friend. There she became the associate principal of a school, and, by her account, whirled in high society. In the summer, she traveled in the Rockies and went overland to California in the company of the United States Cavalry.

Several years later, she decided to move from San Antonio to

Austin, and the peril that found her en route to the state capital would
be the first of many escapades. Thirty miles outside of town, three
masked gunmen attacked her stagecoach. Demanding money from
two wealthy San Antonio men in the coach, they robbed them of several thousand dollars.

Then they turned to her.

"Have you any money?" one of the men asked.

"Yes, a little."

"Then hand it over."

"I shall not," she answered. "How am I to get home without any
money?"

The man held a loaded revolver to her temple.

"I'll blow out your brains," he said.

"Blow away," she replied. "I would as soon be without brains as
without money."

"They gave a ha ha," she recalled, and "disappeared like magic
into the forest."

The other passengers, Taylor reported, "congratulated me on my
firmness."

After a while, she left Austin and began a wider odyssey in the late
1800s. She went to St. Louis, followed by New York City, where she
learned to teach dance. Then Chicago and Chattanooga and Ashe-
ville, where her "reputation grew." Chattanooga again, where she
survived a flood. Charleston, where she kept her "presence of mind"
during an earthquake that crumbled the city. Birmingham, Alabama.
Back home to New York state, where her family said they didn't
understand her. Off to Texas again, where, with money she inherited,
she bought real estate. Finally, she traveled across the Sonoran Desert
to Mexico.

Ultimately, Annie claimed she crossed the continent eight times.
In 1900, either from exhaustion or illness, she entered a northern
Michigan asylum. She had recovered enough by the following year to

teach manners and dancing in Bay City, and there it was, during the summer of 1901, that she sat in a rocking chair, read a newspaper, and experienced an epiphany.

Even with its hazy dates and missing details, Annie Taylor's history of luck and pluck seems to have been accepted by the public. It wasn't easily disputed, and there is evidence that at least some of it was true. Jesse James's gang did hold up several stages in San Antonio after the Civil War. A Mrs. David Taylor did teach in San Antonio, and an Anna E. Taylor did buy an expensive piece of property there. But, above all, it must have been hard to tell a woman who had built a human-sized barrel that she could not have crossed the Mexican desert or stood up to a man with a gun. And if she took that barrel and plunged down the most famous falls in America, she probably could have whatever past she claimed.[5]

III
LADIES OF CONVENTION

On the afternoon of October tenth, about the time Annie Taylor was trying on her barrel for size and fit, the Board of Women Managers of the Pan-American Exposition were buttoning up more conventional attire. They stepped into gowns of silk and crepe de chine and ordered carriages to take them to the Women's Building at the Exposition. There they seated themselves at tables graced with clusters of white carnations, ate ices and cake, and listened to a soloist perform Italian songs. The occasion was the visit of the first lady of New York State, Mrs. Benjamin B. Odell.

If Annie Taylor could have been granted a wish, it would have been to join that company, making polite talk, handling her teaspoon with aplomb, and being complimented on her becoming dress and hat.

She had a way to go. Not only had she arrived at Niagara Falls with a honky-tonk manager and a daredevil scheme, she had also accepted the attentions of the press. Still, she believed she belonged to high society. "I do not want to be classed with women who are seeking notoriety," she asserted. "I am not of the common daredevil sort. I feel refined and I know that I am well educated and well connected."

In Buffalo in the fall of 1901, many of the arbiters of refinement were sitting with Mrs. Odell. Along with the men of their social class, these white women decided who belonged to polite society and who, emphatically, did not. Not all of the twenty-five women whom John Milburn had appointed to the Board of Women Managers were wealthy—some were distinguished through their work in education, the arts, or health professions—but many ranked among Buffalo's upper crust. They led clubs or charities or were known for sophisticated hospitality.

Just as it was elsewhere in America at the turn of the twentieth century, real sisterhood across race and class was in short supply at the Pan-American Exposition. The Women Managers did little public mingling with the Polish, German, or African American women who cooked, served, or cleaned at the fair; the clubwomen of color who gathered separately in the city; the suffragists who rallied in town; the women who entertained at the Midway by dancing or, like Chiquita, charming visitors.

The Board of Women Managers embraced women of Latin America as "honorary" members, but it is unclear how much cross-cultural activity took place. And their beautiful Exposition building—to the dismay of Director-General Buchanan—was designed mostly for private use.

The Women Managers attracted additional criticism by deciding not to feature women's work in a distinct pavilion. This was, they declared, a progressive move. Hadn't women advanced enough, become confident enough, to put their products side by side with the work of men? Besides, in Chicago, visitors had been able to avoid the separate women's display. "Women who attended the World's Fair," explained one writer, "confess to a difficulty in persuading husbands and brothers to enter this sacred precinct."

Others disagreed. "Women are a great deal more interesting as women than they are as impersonal workers," commented one visitor, "and their work is more interesting as women's work than simply as manufactured articles." What this woman did not say, but perhaps implied, was that as much as women might want mixed competitions, they had produced their work in unmixed settings. They lived with assumptions about gender differences. Seen as naturally maternal and domestic, they had fewer professional opportunities, fewer educational options, less money, and arguably less time. Female fairgoers could thus appreciate the obstacles that stood behind everything a woman produced.[6]

However they defined progress, the distinguished women of Buffalo did not embrace the likes of Annie Edson Taylor. To some degree, though, Taylor betokened the future. White middle-class women were becoming more mobile—albeit not as wildly peripatetic as Taylor—but less bounded than ever before. Unlike working women, who had for years crossed cities or worked in fields or stared out the windows of streetcars before dawn and after dark, women of means had remained relatively sequestered. Yes, they went freely to church and club meetings, and visited urban settlement houses, and shopped in special districts, but it was only recently that they had moved about—really traveled—independently.

The "New Woman," as she was known, was a person of education, pluck, and daring, one who wasn't afraid to go places. She was a promising match for the adventurous new outdoorsman, epitomized by Teddy Roosevelt, and she was not-too-subtly encouraged to bear the children of a "manly" sportsman to help sustain the hardihood of the white race. The popularity of this independent woman was under-scored by the sensation of American reporter Nellie Bly, who circled the world alone (in under eighty days) in 1889.

In the continental United States, the New Woman rode trains by herself, giving her even more freedom to cross the country alone than to navigate city streets. She occasionally took the wheel of an automobile, too. At the beginning of September 1901, in fact, before McKinley's assassination derailed the attempt, four women competed in the Auto Endurance Race from New York City to the Pan-American Exposition.

More than anything else, though, the spirited woman of the turn of the century rode bicycles. The invention of the safety bicycle, with wheels of equal size and inflatable tires that cushioned bumps, allowed women to wheel into areas that had once been off limits. Bicycle makers made it possible for women to retain their dignity, too; they dropped the frame so that women could wear their skirts while ped-aling. This bicycle was less structurally strong than a man's bicycle, and often weighed ten pounds more, but that seemed a small price for new freedoms. It was, claimed suffragist Frances Willard, a true "implement of power." Inveterate fairgoer Mabel Barnes frequently traveled by bicycle. On September 28, she arrived at the Exposition after a "wild ride." For a circumspect Victorian like the schoolteacher, this was excitement beyond measure.

And yet limits remained. Women travelers were urged to be cau-tious and careful and warned not to go too far, literally. Female train travelers were expected to use "ladies' cars" and to avoid attention. Women automobile drivers attracted particular suspicion. There were ten thousand cars on the road by 1901, but only a fraction of these were driven by women. Public officials, launching a campaign against women drivers that would carry on for decades, deemed women utterly unsuitable for the road: They couldn't concentrate, they couldn't master car mechanics, they were too emotional. By being on

> Twenty-seventh
>   Visit
> Thursday,
>   October 3, 1901
>   After a
> wild ride on my
> wheel from school,
> I met Miss Hale at
> the East Amherst
> Gate at three P.M.
> We established our-
> selves in the colon-
> nade north of the
> Temple of Music,
> and nearly froze
> while waiting for
> the Carnival Pageant.

Mabel Barnes visits the fair by bicycle.
Excerpt from her Exposition scrapbook.

the loose in cars, they also endangered their virtue. If women persisted
in wanting to drive, they were urged to use electric cars. These were
quieter, and, because they needed to be tethered for power, had limited range. Keeping women off the roads would be a hopeless cause for
the conservatives, however. Even President Roosevelt's oldest daughter, Alice, hungered for the speed and power of gasoline engines and
drove long distances unchaperoned.

Women bicyclists were not immune to criticism. Urged to wear
long skirts to cover their feet, they were ridiculed as "mannish men-
aces." And while some doctors supported moderate bicycling for the

fresh air it provided, other medical men railed against the practice. Riding astride endangered purity and awakened sexual feelings, they argued. Saddles, particularly when the machine was ridden uphill or speedily, encouraged masturbation. Furthermore, white women who persisted in such activity endangered the race. Bicycling stressed women's reproductive organs and reduced the chances for healthy offspring.[7]

The problem with Annie Taylor was that bicycles, cars, and trains—as dubious as they all were for women—were a bit too tame. She would ride a barrel instead, and take it through rapids and over a precipice. And the problem with Mrs. Taylor wasn't simply that she wanted to enter a man's domain. It was that she wanted to master that domain before he did.

IV

THE DEADLY WONDER

Mrs. Taylor arrived in Niagara Falls on October 13, and, with her manager's help, settled into a boardinghouse on First Street, on the American side of the river. Tussie Russell put the barrel on display in a local hotel to drum up interest. For a duo who had boasted about careful planning, Taylor and Russell had left a few things until the last minute. They needed to find someone to navigate the barrel into the correct current. A misjudgment would mean that the cask would tumble over the deadly American falls and strike the massive rock slide at the base of the ledge. And Horseshoe Falls was hardly benign. Who would ensure that Annie moved close to the Canadian shore and captured the safest stream?

Russell hoped that the answer lay with a duck hunter named Fred Truesdale, who, it was said, knew the river's character and currents better than anyone. Officials were not convinced. They suggested she try the Whirlpool Rapids, instead, and reminded her that her stunt would be illegal. People couldn't just kill themselves, they said. Others wondered whether that was perhaps her intention. Back in Bay City, she had sounded desperate. "I might as well be dead as to remain in my present condition," she had told a newsman.

"You don't mean to say in so many words that you are contemplating suicide?" he had inquired.

She had struck back. "Not by any means," she had answered. "I am too good an Episcopalian to do such a thing as that. I believe in a Supreme Ruler and fully realize what self-destruction would mean in the hereafter."

And in Niagara Falls, when the mayor of the city had asked her whether she was "the fool that's going over the falls," she was taken aback. "The fact is," she said, "I find people are very suspicious when one tells of going over the falls."

She reminded everyone of her barrel's scientific construction: It contained cushions to shield her head and body from the blow of the descent; it offered handholds to give her a grip, no matter which way was up; it had a new harness to prevent the top of the cask from crushing her skull; it would contain a flask.[8]

Only one thing, she admitted, loosened her nerve, and that was the river itself. She remembered it well—she said she had visited Niagara Falls plenty of times while growing up. But she did not want to inspect the scene now. Had she actually taken another look, she might have paused. The thirty-seven-mile-long Niagara River, which separated New York from Ontario, put on a deceptive show. Flowing easily from the east end of Lake Erie, it ran quietly north, almost on a level with its banks. Its current was mild, and, at five miles in, it divided around Grand Island.

After converging again north of Grand Island, the river began to change character, and, about three-quarters of a mile above the falls, it quickened its pace and began to move furiously around rocks. As it approached the grand precipice, it divided again, this time around little Goat Island, and then it descended 170 feet over Horseshoe Falls on the Canadian side and a shorter but more treacherous distance over the wider, rockier American cataract.

It was the water's sensuous curve over the edge of Horseshoe Falls that captivated visitors. Produced by rock salts and pulverized lime-stone, the glasslike green was so entrancing to Pan-American Exposition designers that they tried to add a touch of it to almost every big Exposition building. If a hall was painted a primary red, it would still be trimmed in Niagara green; if its walls were ivory, it would still sport the wondrous translucence of the waterfall.

After cascading down into a pool, the water of the cataract moved through a high-cliffed gorge, creating the rapids that Martha Wagen-fuhrer conquered and the Whirlpool that caught Maud Willard in its death spiral. Then, placid again, the river bent north toward Lake Ontario.

Annie Taylor came to Niagara Falls in 1901 because the Exposition crowds were there. And the crowds came because they wanted to see one of the world's natural wonders before it shifted shape. At the turn of the twentieth century, they were told, erosion was moving the edge of the falls upstream toward Lake Erie at the rate of at least four feet per year. The diversion of river water to make electricity slowed the erosion but endangered the spectacle. Either way, it was time to visit.[9]

Back in the early nineteenth century, tourists had come to Niagara Falls with less urgency. The completion of the Erie Canal in 1825 gave travelers easier access to the area, and artists, writers, and honeymooners flocked to the site. By 1857, when landscape painter Frederic Edwin Church completed his *Niagara*, a seven-foot-wide work that drew thousands into exhibit halls in the United States and Europe, more than sixty thousand people a year were visiting the area. By the turn of the twentieth century, after the completion of more bridges, restaurants, hotels, and shops—not to mention hydroelectric plants—excursionists to the falls numbered over three hundred thousand a year.

Then there were the performers. The heyday of stunters began in 1859, when a Frenchman named Jean-François Gravelet, known simply as Blondin, stretched a rope two thousand feet across the Niagara Gorge and crossed over on it. At the halfway point, he lowered a bottle to a tour boat 160 feet below, waited while it was filled with river water, then pulled it up and drank it. He walked the tightrope again and again that summer. And he didn't just walk it. He carried his manager across it, danced on it, and did handstands and somersaults. He traversed it blindfolded and in the moonlight. He took a stove across it, lit a fire, cracked eggs, and cooked an omelet. He took Champagne and cake.

Blondin became a sensation, and he attracted others who tried to outdo him, including a man dressed as a washerwoman who scrubbed laundry in a tub while on the tightrope and a woman who crossed the gorge with peach baskets on her feet. Every one of the fifteen tightrope performers survived.

By the mid-1890s, though, tightropes were passé and daredevils chose to prove their mettle in the Whirlpool Rapids, where the river flew by at forty miles an hour. Matthew Webb, a champion of the English Channel, was one of the first to try the rapids. He arrived in Niagara Falls in 1883, entered the river, and drowned. Others challenged the rapids with rowboats and barrels, and some of them, like Martha Wagenfuhrer, lived to tell their tales.[10]

If gut-wrenching horror was what crowds wanted at the falls, then two young men—an Englishman named Arthur Midleigh and a Canadian named Alonzo Gardner—topped it all in 1889. Their attempt to row across the rapids above the cataract was, said a reporter later, "the most awful and thrilling catastrophe that has ever happened at that scene." The men misjudged the strength of the current, and, 250 feet from the brink of Horseshoe Falls, they scrambled onto a rock while their boat tumbled over the edge. People, including Gardner's wife, gathered on shore to watch. Rescuers sent a boat to the rock, tethered to the shore. The boat smashed. Then they maneuvered a wooden raft to the beleaguered men. Midleigh leapt onto it, lost his footing, and fell into the water. Spectators, more than a thousand by now, watched his head rise above the water just before he went over the precipice.

The next day, tourist trains to the falls were full. Twenty thousand people gathered to see Alonzo Gardner be rescued or die.

Men rigged up another contraption. Hoping to secure Gardner against the powerful current, they crafted a harness out of a hotel fire escape, with a hook tethering it to another jury-rigged raft. They delivered it to the rock. Sluggishly, awkwardly, Gardner strapped it on. With no rest for three days, he seemed to move in slow motion.

Finally came the moment the watchers on shore dreaded. Gardner had to hook the raft. At first try, it launched out of the river and flew past him. He steadied himself. Once more, the raft shot up out of the froth. Then suddenly, in a second that must have burned forever in the minds of those looking on, the raft spun and knocked Gardner into the river. He went down, said a witness, like a bowling pin.

Like his friend before him, Alonzo Gardner was carried toward the edge. At the end, he raised his arms, and, in a turn of the current, seemed to give one long look at all the people on shore before he went over to his death.

It was events like this that discouraged daredevils. Instead, they used animals. In the late 1800s, dogs, cats, and even a rooster were sent over the cataract, with and without barrels. Some of the animals perished, and some, like Frank Bostock's crocodile Ptolemy, were pulled out of the pool below the falls in shock, but still breathing.

Annie Taylor also sent an animal over the falls. On Saturday, October 19, the river man she had hired, Fred Truesdale, who had previously experimented with animals, caught a cat, placed it in Taylor's barrel, and let it loose two miles above the falls. To his great satisfaction, the barrel plunged over the falls bottom first and appeared, undamaged, a minute and a half later. The captain of the *Maid of the Mist* tour boat pulled it out of the water. The cat reportedly jumped out, very much alive. Yet witnesses disagreed. One said the animal had suffocated. Another said it had drowned in water at the bottom of the barrel. He had seen the body.[11]

V

THE REAL THING

It was deep fall now at the Pan-American grounds, and the cool days and colder nights had a withering effect. Flower beds, once the pride of the formal courts, had yellowed and dried. Displays that had been resplendent with color only weeks earlier, had wilted. Even the asphalt was stained and spotted, where dung, grease, and human traffic had left their marks.

On the Midway, some shows, like Bostock's, happily counted their cash, while others contemplated a shortfall. The Streets of Mexico, nearly bankrupt, had been forced to close early. The electric lights at the popular Alt Nurnberg Restaurant were temporarily shut off by the Exposition company, which claimed that the restaurant owed $8,000. The Esquimaux Village concessionaires, on the other hand, wanted money back: The Alaska exhibit had copied some of their acts. Darkest Africa didn't have a grievance but just needed to pull out early: Its performers had to catch a steamer to Europe.

Even Director-General William Buchanan was leaving. Now considered an expert in Pan-Americanism, he was heading to Mexico City for a meeting of the Pan-American Congress, so he left others to close up the fair. Exposition officials, honoring him with a farewell dinner on October 24, produced a large Tiffany loving cup for his two years' hard work, and concessionaires bestowed on him a diamond ring—a "token of his kindnesses and fair treatment."

Newspapers now became boldly pessimistic. On October 22, the *Commercial* said it outright: The Exposition was a financial "failure." And it went further. Given "the beauties of the fair and the abundance of attractions round about Buffalo, the total number of admissions should have been 15,000,000." The editor blamed the low admissions on the railroad rates, but wondered whether people were weary of fairs. It wasn't just Buffalo—nobody could put on a big show anymore.

The *Express* demurred. Wait, it said. The fair would go out with glory. Farewell Day, November 2, would be an eye-popping spectacle guaranteed to fill the Exposition's empty purse.

Bostock, as ever, seemed oblivious to the pessimism. Throughout October, he was a figure of animation and good spirits. In the third week of the month, he admitted thousands of children into his show for free. And he continued to advertise new animals. Trains brought in an emu, a group of Chilean armadillos, some cassowaries, and a family of baboons. The menagerist encouraged crowds to visit his celebrated acts, too, before the show closed. He made sure that Chiquita was as beguiling as ever in these final weeks, and he sent Jumbo II into Exposition parades, most recently plodding alongside a procession of fancily dressed babies.[12]

It was during the third week of October that rumors out of Niagara Falls gathered steam. Soon enough, they had the heft of a news story and their strange, startling content began to captivate the country.

Annie Taylor seemed serious about sending herself over Niagara Falls. As her plans materialized, reporters started following her from place to place, and, as they did so, they scrutinized her more thoroughly. There were some things about her that did not seem right. Besides the obvious madness of her barrel ride, she didn't look the part of a daredevil. She was a dancing teacher, she said, and an instructor in physical culture. Yet she was a woman of sturdy proportions. One paper estimated her height and weight to be five feet four and 160 pounds and went so far as to declare her "quite stout."

Taylor explained that she was more flexible than she seemed. She was, she said, "able to bend forward and touch her nose to the floor without bending her knees with perfect ease." She said she could turn cartwheels, and, up to two years earlier, she could "draw herself up on a trapeze and hang by her toes." She did not demonstrate. She was, she explained, "out of practice."

She didn't tell them the truth. She billed herself as forty-three years old.

In fact, she was sixty-three.[13]

Taylor told newspapers that she would make the plunge on Sunday afternoon, October 20. But the afternoon came and went. Watchers gathered and grew weary. By Monday, tourists and voyeurs and newsmen were becoming impatient. One reporter wondered whether the whole thing was a "gigantic hoax."

Taylor and Russell offered excuses. They were having a few difficulties with the authorities—both Canadian and American—who had had the temerity to say they wouldn't permit the life-threatening stunt. Her publicity was also unready. Taylor wanted to sell photos at the time of the descent, but her photographer had been delayed by a fire near his studio.

She bided the time with more interviews. She listed her barrel equipment again for skeptics and mentioned the weighted anvil at the foot of the cask that would help her land feetfirst. And she admitted some doubt. There was a possibility, she said, that the fall would "break my neck." She also acknowledged that, in the dark of the barrel, she wouldn't have any idea where she was heading. She was sure that she would know when the barrel went over the edge, though. And, she said, "I am quite positive I shall know when it strikes the water below."

Indeed.

The day she promised for her stunt, Wednesday, October 23, opened sunny. The wind blew in gusts, however, and threw the fast-moving Niagara River into a chop. Sightseers crowded onto the shores by early afternoon, and reporters gathered over at Fred Truesdale's house on a shallow inlet known as Port Day. Hoping to watch Taylor get into her barrel in the Truesdale woodshed, they were disappointed. Concerned about propriety, she crawled in privately, took a swig of brandy from her flask, and tucked it in beside her. She then welcomed photographers to see her, or at least her head, inside the barrel. The press was pleased. They wanted to be sure it wouldn't be a con job.

Using long crowbars, four men carried the heavy barrel down to Truesdale's dock and maneuvered it into his boat. A policeman appeared on the street above the house. The men hurried their work, said quick good-byes, and Truesdale and an assistant named Fred Robinson rowed the barrel from the shore.

The policeman shouted. Too late.

Meanwhile, gusts on the river had increased, forcing the men to row their cargo into a headwind, water spraying over the gunwales. Worried that their boat might founder, and that they might go over the falls with their passenger, the men beached the boat at the reedy outcropping of Grass Island.

From inside the barrel, Mrs. Taylor urged the men to proceed. The boatmen said they would not. She asked them to at least wait to see if the wind calmed down. They did, yet it persisted. Finally, she gave up, too. The boat's intense rocking, the tilt of the barrel, and the close, dank smell began to make Annie sick to her stomach. She got out. And, sitting in the bow of the boat like an oversize and forlorn figurehead, she returned to shore.

To the crowds who had patiently waited, and who now began talking about "cranks and fakers," she offered determined words. "I will make the trip tomorrow," she said. "I think the people are convinced by this time that I do not intend to deceive them. I hate a weak, vacillating person whose yes means no and whose word cannot be depended upon." Another report said she was so unhappy about disappointing people that she cried.[14]

The next day, Thursday, October 24, was a good-luck day. Or it would make for a nice, round ending. It was Annie Taylor's birthday. She did not bring up the topic, of course. There was no need to fuel speculation.

The winds had died overnight, and the weather was clear and cold.

Hoping to avoid the police, who anticipated a late-afternoon event, Fred Truesdale decided to leave shore with the barrel around 1:30 p.m. Meanwhile, Tussie Russell allowed newsmen a last-minute meeting with Annie. "I believe that I can live fully an hour, perhaps two, with the cover closed," she said. Once again, she admitted some concern. "I have a terror of the Whirlpool Rapids, and I do not want my barrel to escape the watchers below the falls and get into the Whirlpool."

An hour passed. The boatman who had helped Truesdale the previous day was missing. Men carried the barrel down to Truesdale's sailboat, and waited. One thirty. Two o'clock. The police were surely on their way. Reporters took up the slack. One from the *Express* asked Mrs. Taylor whether she had considered that this might be her last living hour. "I have made all arrangements," she admitted. "I have told Mr. Russell to telegraph to my sister . . . and notify her of my death if I should not survive."

The boatman arrived, explained his delay, and said he had bad news. The police had told him he was assisting someone with suicide, and he was scared. "I don't want to get pinched," he said. He was out. Fred Truesdale, acting fast, turned to Billy Holleran, a young fisherman experienced on the upper river, and pressed him to join him. Holleran agreed.

By the time Mrs. Taylor emerged from the Truesdale house and headed to the dock, she had to make her way through a channel of spectators. Peter Nissen, a stunter who had rowed a boat through the Whirlpool Rapids less than two weeks earlier, wanted to shake her hand.

"You are going to beat us all," he told her with a smile.

"Oh, I don't know," she replied, "although I am going to try."

She was helped to a seat in the boat's bow—she would enter the barrel at Grass Island this time—and, just before shoving off, she turned to a reporter and handed him an empty envelope with her sister's address. "Will you let her know?" she asked. "In case of accident and I should not come back?"

It was the surest sign yet that Annie Taylor knew she was likely minutes, an hour at the most, from dying.

But, just as quickly, the bravado was back.

"Good-bye, Mrs. Taylor," the reporter said, as her boat began to move.

"I won't say good-bye," she replied. "It's just au revoir, you know. I will see you all again."

She was pulled away from the American shore.

Past a breakwater to Grass Island, a flotilla of photographers, reporters, and boatmen followed her. The water barely registered a ripple.

Her barrel was rolled into the reeds, and, looking ready for business in a dark tailored jacket, she posed for photographs. Meanwhile, noticing that the barrel had developed a small leak, Truesdale applied some caulk.

The men moved to the other side of the island while Taylor took

Annie Taylor inside the barrel.

off her jacket and overskirt and loosened her shirt. She pulled off her hat. The tall reeds on the island concealed her as she levered her heavy form into the barrel.

Truesdale and the other men returned. They slid the lid shut and turned a wooden screw. The next time it would be opened was when it was all over. From the inside, Taylor uttered a muffled shout that there was another leak. She could see daylight. Truesdale fixed it.

Holleran worked a bicycle pump to put fresh air into the cask. When the pump began to work hard, Truesdale asked whether she still saw light. She did not. He told her he thought she had enough air.

The men rolled the barrel into deep water and tied it to the stern of Truesdale's boat. They rowed to Hog Island, near the mouth of Chippewa Creek, brought the boat about one-quarter mile above the head of the rapids, and drew the barrel in close. There would be no returning now.

Inside the tight, black space, trying to take in the pumped air, Annie Taylor couldn't breathe well. The air seemed thick. She wanted to try the breathing tube over her head. She needed to know that it worked. If she wasn't found fast enough, she would have to have more air. If Maud Willard ever haunted anyone after her grim, suffocated death, it was right now. Taylor did not want the Whirlpool.

She opened the cork. Water came through, not air, and she gulped it in. This wasn't right. Just then, though, she heard the rapping. Truesdale's oar banged on the barrel. She had been cut loose.

"There is water coming in here," she tried to yell.

The men heard her—a small voice now, deep inside the wood. But they were done with fixing.

They let her go. Even then, they barely had enough time to save themselves. "For a moment or so," recalled a witness, "the boat swung along in the current, moving faster [downstream] than the barrel, and then Truesdale and Holleran took up their oars." They got away.

It was four o'clock.

Along the Niagara River, and at every precipice and island and fence and overhang, people crowded the shore. A few of them had raced from Port Day in order to see both the beginning and the end. They watched from Goat Island, Three Sisters Island, Terrapin Point, and Table Rock. Some had been there with their binoculars for hours. Some had been there the day before.

The watchers saw the barrel drift steadily toward the breakers, hit the whitewater, get pitched and tossed, disappear from sight, then reappear. It shot into the air. It looked like it was hitting spray around rocks, but it made headway, spinning and bouncing through the rocky terraces, going lower, still intact.

Bit by bit, the barrel moved toward the brink. Suddenly it caught, as the anvil lodged against a rock. The crowd gasped. Just as suddenly, it loosened and careened onward. Mrs. Taylor's death, it seemed, would be fast and soon.

As the barrel inched closer to the edge of the falls, it followed the curve of the shore and entered the smooth water—a gliding sheet of green. Niagara green. The honored color of the summer. It was a beautiful sight—the inexorable, unstoppable power, and the clear, verdant color.

As the barrel approached the edge, onlookers began to work feverishly at their cameras. They tried to point their lenses, but the dark shape moved fast.

Mrs. Taylor reached the precipice. And tumbled over. The cask fell bottom first, straight down.

"She is over!" shouted the watchers, in unison.

Then the barrel disappeared.

Below Horseshoe Falls, Richard Carter, captain of *Maid of the Mist*, circled in the froth. He wasn't sure what to look for: pieces of wood, a barrel, or, just as likely, if the force of the falls had sucked the cask into its lowest vortex, nothing at all.

Then, there it was, in the current along the Canadian shore.

He blasted his whistle.

A rescue party, including the ubiquitous Carlisle Graham, stood on a rock ready to help. The barrel moved into Bass Rock eddy, about a quarter of a mile from the falls, and then edged closer to shore. A man stripped down to his trunks, seized the barrel's handle, and, with help, pulled it to safety.

Up above and along the banks of the gorge, people cheered.

Others knew it was too soon to cheer. The lid had to be opened first.

Three men worked to unscrew the lid and then slid off the oval top.

They looked in.

A white, water-soaked face. A blue hand.

The hand extended through the opening.

And waved.[15]

9.

The Escape of the Doll Lady

I
BRAIN FEVER

Annie Taylor was still alive—that much was certain. Soon enough, though, it became apparent that she was stuck in the barrel. Waterlogged, cold, and weak, she couldn't edge herself out, and nobody else could either. The thick oak that had shielded her from the blows of a thousand rocks now encased her like a form-fitting coffin. It was unclear how long she could last, or what shape she was in.

The men sent for a saw. With that in hand, they cut their way through the oak, took off the top of the cask, and, with a mighty tug, pulled her out. She had a bleeding laceration behind her ear and she looked shipwrecked. But, with help, she stood up and spoke.

"Have I gone over the falls?"

Annie Taylor's mind was muddled, but her limbs were intact. She walked upright along the rocks and climbed across a plank to a boat. Photographers captured the moment—her hair in a bedraggled topknot, her dress disheveled, her hand outstretched for help. Within an hour, she was back at her boardinghouse, sipping coffee and brandy and thawing out with the help of a stack of blankets, hot-water bottles, and a coal fire.

When she felt well enough, she spoke about what she remembered. As she plunged through the rapids, she said, she thought she was suffocating. But worse, she recalled, was the noise that came rumbling through the barrel's thick walls. It was the deadly roar of the falls, and it grew to a thunder.

She felt a short drop. Rocks, she imagined. Still the rapids.

Annie Taylor, dazed and triumphant.

Then came the "terrible nightmare." She couldn't say whether the barrel had stalled, or risen up, or lurched. But all at once she felt hollow, and her stomach flew into her throat. "Something," she said, "had given out from under me."

The next thing she knew was that she hit a rock, slamming into it so hard that her barrel broke and cold water struck her face. Later, of course, she realized that it wasn't a rock at all, just the impact, 158 feet down. Her barrel hadn't broken, either, just leaked. The water, she remembered, felt "so cold."

She thanked people, like Captain Billy Johnson, who had urged her to install arm straps. He had helped her avoid "having my brains beaten out." And she thanked God. "I owe a debt of great gratitude to Him," she declared.

Asked whether she would ever do it again, she had a quick reply: "I'd sooner be shot by a cannon or lose a million dollars than do it again. I will never do it again. But I am not sorry I did it, if it will help me financially."

Tussie Russell was already at work on that score. Annie Taylor was receiving offers of marriage, paid interviews, and appearances. Most urgently, there were offers from the Pan-American Exposition, and he was trying to hold out for a lot of money. He also had his hands full with the barrel itself. Even before the night was over, souvenir hunters had broken up the hatch, cut the harness to pieces, and hustled them away. Maybe they wanted a part of history, but, more likely, they hoped to sell it.

There was one hitch in the plans, however. Annie's stunt began to catch up with her. Bruises had begun to mottle her body, and she felt shaky—even "hysterical." The next day, when she still felt unwell, a doctor did an examination. He saw the signs of a frightening illness— brain fever.[1]

II
THE BLIND SPOT

While Taylor lay sick and Russell fretted, another press agent, on a different account, stepped up his sales pitch. In the third week of October, Captain Jack Maitland released a storm of stories. Time was running out, he said, to go to one of the most remarkable performances in the whole United States. No one should miss Jack Bonavita's twenty-seven lions, or Madame Morelli's jaguars, or the "funny clown elephants." There was a new show, too, just for "windup week." Madame Hindoo was in town, with her serpent charmer Princess Brandea and their deadly cobras.

The publicist also let it be known that Frank Bostock was being honored. "New laurels have been added to Frank C. Bostock's reputation," Maitland asserted. The agent did not specify what laurels had been awarded or by whom. And he ignored the fact that in an end-of-the-fair contest for Mayor of the Midway, four concession heads had been nominated: Frederick Cummins of the Indian Congress, Frederick Thompson of "A Trip to the Moon," Gaston Akoun of "The Streets of Cairo," and Fritz Mueller of "Pabst in the Midway." Bostock, despite his spectacular profile—or perhaps because of it—was not in the running.

Captain Maitland also made a final push for the Doll Lady. Chiquita, he announced, had been a stupendous success, and the parlors of the Exposition mascot had been thronged with visitors. No one should pass up the chance to meet the refined little person who had mingled in the highest society. Taking the public into his confidence, Maitland even whispered a little secret about the human sensation. Chiquita, he disclosed, needed a lot of sleep. "She rests ten hours at a stretch."

Alice Cenda probably disliked having her personal habits shared with the public, but this might have been an exception. In just a few days, she would use this information as a cover. She and Tony, in their fleeting meetings, through go-betweens and in notes, had worked out a plan. On the second-to-last night of the fair, she would pretend to go to bed. Hours later, after her attendant and Bostock himself had retired, Tony would help her break out of her prison.

They thought they had worked out a good escape route. The Animal King bolted Chiquita in her rooms every night, and he had done a good job of securing her windows and doors. But he had missed something. Below Chiquita's ticket window, customers had left a depression in the ground. This depression, if deepened, could provide an opening.

The performer bided her time. She did her best to please her manager, and, just as she had for months, she carried out her act seamlessly. She showed off her languages, signed autographs with her small, round script, and greeted hundreds of curious visitors. Back in the summer, she had had to cancel her show because shaking hands had hurt her. Now she allowed visitors to squeeze her small palm, minute after minute, all day long, without complaint.[2]

III
RICH PEOPLE, POOR PEOPLE

The last days of October at the Pan-American Exposition, just before the grand finale, hummed with purpose. It was a time for dismantling and packing and making unashamed efforts to squeeze the last pennies from fairgoers. Salesmen marketed everything from portable souvenirs to big buildings and their contents. It was a time of giving gifts for jobs well done. Concessionaires handed their director, Frederick Taylor, a gold-headed cane; garden workers gave their chief a silver punch bowl; and the electrical engineers presented their head with an inscribed gold watch. Not to be outdone, the Exposition Hospital staff honored their superintendent, Adele Walters, with a silver-topped umbrella and gave the long-suffering Exposition physicians new instrument cases.

It was a time, too, for hearsay. Stories coursed through Buffalo and the fairgrounds. Everybody agreed that the Exposition cost a lot more than expected and attendance was a lot less than they wanted, but what did this mean? Who would be out of luck? Who might not be paid? And, above all, could the fair do better? Seven and a half million tickets had been sold so far. Was there hope for more?

Buffalo businesses were urged to let their employees have one last look at Rainbow City, and bring up its numbers. Any day in the next six days would do, promoters said, just—please—boost the attendance at the "greatest enterprise ever planned and carried out by Buffalo." If nothing else, they said, go on Farewell Day, Saturday, November second. And, as though they couldn't help it, they pulled out their timeworn tactic: Shame people with the White City. "It should be the aim of everyone to swell the Farewell Day crowd, to make it reach the 200,000 mark," wrote the *Buffalo Commercial*. "At Chicago a daily attendance of 300,000 was not uncommon toward the last of the exposition season."[3]

Pan-American directors advertised energetically these last days, but they didn't want just any visitors; they wanted paying customers. It was brought to their attention, however, that some people in western New York hadn't been able to afford the Exposition. The parents of some local schoolchildren had found it a challenge to scrape together the round-trip streetcar fare of six cents, much less the discounted admission ticket of fifteen cents.

Learning of this, Buffalo philanthropists rose to the occasion and bought trolley and admission tickets for six thousand children. Impoverished adults were a different matter. A rooming-house owner sent an appeal to the press, explaining that three of her lodgers had never been inside the fairgrounds, though "how much they would like to go" she couldn't begin to say. And why had they missed out? "For the simple reason that they cannot afford it." One, she said, was a woman and her young daughter, who both worked hard. The mother washed and cleaned houses, and some days she could hardly crawl home. But "they do not have one penny to spend—hardly enough to buy clothing." Another was a woman with a baby whose husband had left her, and the third was just dirt poor. "Why not," she pleaded, "let the poor people of Buffalo . . . see a sight of their lifetime?" Why not let them have "one day of recreation in which to forget their poverty?"

The answer, not surprisingly, was *no*. For an exposition company that could not at that moment pay its contractors and builders—not to mention mortgage holders—generosity, or justice, was out of the question.

To paying visitors, Pan-American promoters offered an unforgettable final day. First of all, fairgoers would be able to leave a legacy. They could drop their names and addresses into boxes on the grounds, and a record of their attendance would be stored in the fireproof archives of the Buffalo Historical Society, where it would be kept for "all time." (It wasn't.)

In addition, the fair's famous Indian warriors would kill Custer once and for all, and then attack a wagon train. The United States Cavalry would then avenge Custer. Indian women would be making food, the men would be resting, and suddenly the soldiers would strike. It wouldn't be the usual "harmless" battle, either. The 150 Indian fighters, being real soldiers, and 300 Indians, being real Indians, meant that the "shooting and dying and scalping will be done in truly a realistic way."

This fight, one newspaper said, would sum up the story of the fair. "Pitted against each other," the *Commercial* explained, "will be the representatives of the forces that have met in deadly conflict for over two centuries in the New World—one protecting the advance

The popular sham battle between the Indian Congress
and United States soldiers, in the Exposition stadium.

of civilization, the other virtually courting extermination in its efforts
to retain a continent for primitive savagery." Since the nineteenth
century had seen the triumph of civilization, and the Pan-American
Exposition had celebrated that triumph, it was only fitting that at the
end of the fair these two groups should engage in final, "friendly"
conflict.

If this reporter had paid some attention to the news items his
colleagues had produced over the Exposition season, he would
have realized his summary left a few gaps. True, the local press
had devoted countless inches to describing the backwardness of the
Exposition's indigenous peoples. But, intentionally or not, it had

added nuance to the picture. It had described how Geronimo, for instance, had been a tireless entrepreneur, promoting the sale of crafts and negotiating contracts. It had recounted the ways he had defied Exposition visitors and unapologetically endorsed the dog feasts. It had depicted other Native performers laughing at white audiences, turning the tables publicly and secretly, in ways modern and not modern, embracing "progressive" and "primitive" both. And it had revealed the ways that the Native performers had established dynamic Exposition communities. The fair served as a great inter-tribal powwow, with dances and feasts that cemented alliances and showcased resilience.

The Indian Congress show would lead off the events on Farewell Day, but there would be other big acts. "Brawny" Irishmen would oppose each other in a hurling match, and on the Midway, cakewalk-ers, pie-eaters, and spielers would vie for prizes. The committee for Farewell Day would bring in special celebrities, too. Carrie Nation would draw crowds with her inflammatory temperance talk, and word had it that barrel rider Annie Edson Taylor had (mostly) recovered from her descent over Niagara Falls, and would be willing—perhaps even eager—to receive visitors at the Exposition.[4]

IV
DEADLY FORCE

As they had done for most of the fall, fair directors and city leaders stirred up excitement for the Exposition even as they beat back memo-ries of September. They had even, begrudgingly, worked assassination scenes into tourist itineraries. The Temple of Music, the Exposition Hospital, the Roosevelt inauguration site, and the Milburn house had become must-see locations.

On October 29, though, the painful drama of early autumn once again took over the news. It was the day designated for Leon Czolgosz's electrocution at the state prison in Auburn—four days ahead of the fair's grand finale. The papers were full of the event, and no matter what dazzling shows had been conjured up for the close of the fair, this story was impossible to ignore.

Among the men invited to witness the electrocution was Charles R. Skinner, who had served in Congress with William McKinley and was now an official with the New York legislature. He was pleased to be at Auburn. He had been unable to attend the execution of Charles Guiteau—President Garfield's assassin—because of a daughter's illness, and he did not want to miss this one. He arrived in Auburn the night of the October 28, visited with friends, and ordered a wake-up call for six in the morning.

Inside the prison, meanwhile, two physicians made a final examination of Czolgosz, and again convinced themselves of his sanity. They were so impressed by his mental faculties, in fact, that one asserted that he seemed "exceptionally intelligent for one in his walk of life."

The next day, a Tuesday, was cloudy. The weather was just right, Skinner felt, for what was to come. In the early morning, he presented his "invitation" to the prison warden and gathered with thirteen other witnesses in the warden's office. They were led down a corridor to the death chamber.

Except for witness chairs and the prisoner's oak seat, the room was bare. There were windows, but they were too high to allow anyone to look in or to offer the condemned a glance outside. At the back of one wall stood a small enclosed area—the "executioner's box." Inside, modern technology was manifest: electric lights, electric wires, an electric bell to inform the occupant when to pull the switch. The switch itself was brass, with an insulated handle. The man who closed the electrical circuit could not see the prisoner. Nor could the prisoner see him.

It was just after seven o'clock when the witnesses heard shuffling in the corridor and a lock clicking open. They saw Czolgosz, between guards, enter the room and walk to the chair. He was expressionless. Guards tied the restraining straps around him and attached the headgear.

The prisoner was then permitted to speak.

"I killed the president because he was the enemy of the good people, the good working people," he said. "I am not sorry for my crime. I am sorry I could not see my father."

Inside the box, out of sight, the electrician—a man named Davis—waited for a signal from the warden. He got it. Then, for just over a minute, and at intervals, he delivered varying voltages—seventeen hundred being the maximum—into the body of Czolgosz. Davis was satisfied with the result. "It was," he said, "as successful an execution as I have ever operated at in all my experience." Other attendees concurred. There was "no scene," one commented. "Everyone conducted himself with remarkable sang-froid."

Like the others, Skinner watched the extinguishing of life with little emotion. There was no more terror in the scene, he asserted, "than to see a cat catch a rat." The prisoner's face had been obscured by a black cap, and Skinner and his fellow witnesses had simply focused their minds on Czolgosz's despicable act. "We all thought of the great crime against our country," he said, "and nothing of the poor form in the chair."

Dr. Carlos MacDonald, an attending psychiatrist, was not as indifferent. He took pains to note that Czolgosz's body was "thrown into a state of extreme rigidity," as soon as the circuit closed, and that every muscle had gone into a "tonic spasm." Another witness, a sheriff, noted that after the first shock, when physicians felt at the jugular vein for a heartbeat, they asked for a second round of electricity.

Physicians performed the autopsy almost immediately. To their relief, Czolgosz's brain seemed perfectly normal. Examining it in full, they could see nothing that spoke of pathology. They thus concurred with official psychiatrists and prosecuting attorneys that the assassin had been sane.

Later that day, what remained of Leon Czolgosz was placed in a plain black casket, doused with sulfuric acid and quicklime to encourage disintegration, and buried. Czolgosz's brother, Waldeck, who had traveled to the prison from Ohio, paid a visit to the grave. Speaking to the press, he said that he had no plans to change his name or to go into hiding.[5]

And so it happened that, in the fall of 1901, the Pan-American Exposition, whose designers had worked so hard to demonstrate all the beautiful things electricity could do, became suddenly linked with the way electricity could kill. For months, Rainbow City had dazzled the public with art and sculpture that celebrated the hydroelectric power of Niagara Falls. It had captivated visitors with the magic of transformers. Its magnificent rheostat, mirroring the coming of dawn and dusk, had taken the spectacular lights of Chicago's White City and moved them, quite literally, a step ahead. Its electric streetcars had taken visitors to and from the fair and its streetlights had illuminated walkways. Its electric elevators had ascended the fair's highest tower, named in honor of this wondrous technology. The Queen City, some declared, should be known from then on as Electric City. Now, though, the public was reminded that electrical current could inspire fear as well as awe.

Rainbow City's founders and fairgoers had not, of course, been unaware of electricity's dangers. Stories of fatal accidents near power stations and power lines appeared in the news with regularity. In New York State, in two recent years alone, ninety people had died by electric shock. By 1901, American consumers were so apprehensive about using electrical appliances that many of them shied away from newly patented toasters and irons. And at the Pan-American fair, electricity was blamed for a peculiar malady. In October, the *Buffalo Express* printed a column about a newly discovered ailment, the "Exposition Collapse." Its symptoms were "exhaustive nervousness" and, in the worst cases, "nervous prostration." Apparently, said a consulting physician, people believed that the "continuous use of such tremendous quantities of electricity, creating such a powerful light night after night for six months had resulted in diminishing certain properties in the atmosphere, whose presence was beneficial to the nervous system." The physician readily dismissed the association, arguing that nervous prostration in Buffalo was more likely due to the "steady strain of receiving guests and sightseeing and rushing around."

The public was also mindful of the dangers of electricity, thanks to the high-profile rivalry between Nikola Tesla and George Westinghouse and their support for alternating current on the one hand, and Thomas Edison and his low-voltage direct current on the other. In the early 1890s, Edison threw money and celebrity into persuading the public that alternating current was hazardous, and he demonstrated its deadliness by using it to kill animals, mostly dogs. He could do little, however, to explain away the fact that alternating current could travel long distances, and was less costly.

Buffalonians had certainly read about the "current wars." If they knew anything about recent history, they were also aware that their city had a unique relationship with electrocution. One of the inventors of the electric chair was a Buffalo dentist named Alfred Southwick. In the late 1880s, he launched a campaign for more "humane" capital punishment. Representing the bluntly named New York State Death Commission, Southwick wrote in 1887 to Thomas Edison, asking for advice on electrocuting the condemned. The commission wanted an alternative to hanging—something more "civilized."

Edison replied, saying that he couldn't help. After a few months' consideration, however, he wrote back, saying he had just the thing: one of George Westinghouse's machines. "The most suitable apparatus for the purpose," he advised, "is that class of dynamo-electric machine which employs intermittent currents. The most effective of those are known as 'alternating machines,' manufactured principally in this country by Mr. Geo. Westinghouse." When the current from these machines entered the human body, Edison explained, the result was "instantaneous death."

The commission was impressed and urged the New York State legislature to move forward. They did, setting 1889 as the date by which capital punishment would make the transition. Edison's men did not hide their hope that AC would become known as "the executioner's current," and that condemned men would be "Westinghoused" in the same way that they had been guillotined (after French physician Joseph-Ignace Guillotin).

Westinghouse, outraged, tried to hire an attorney to challenge the constitutionality of "electricide." When William Kemmler, a vegetable peddler from Buffalo, became the first person condemned to die in the electric chair, Westinghouse paid to help Kemmler appeal his death sentence. The appeal failed. To make matters worse, the protracted electrocution of Kemmler in 1890 in Auburn was nothing short of a disaster. The prisoner shouted in pain, bled, gasped; his skin and hair burned; and his coat caught on fire. He had, said some, been "roasted to death."

Eleven years had passed since Kemmler's death. Alternating current not only had become standard for executions but also was used by industry nationwide. Edison's campaign had backfired. George Westinghouse had been awarded the contract to light Chicago's White City, and, at that fair, as well as in Buffalo, the technology linked utility with spectacle and awe.

Events in the fall of 1901 brought the perversities of electricity to center stage, but not everyone was disgusted by them. If for some people electricity linked distressing images of beauty and death, for others the association was strangely compelling. In this era of science and reason, individuals were sometimes drawn to mysterious forces—they went to Niagara Falls, after all—to see how close they could come to fatality. The electric chair held this same sort of dark appeal. It helped explain why more than a thousand people applied to see Czolgosz's death. And it helped explain why, shortly after that electrocution, hundreds of people filed into Rainbow City to see the Animal King attach electrodes to the ears of an elephant.[6]

V

THE WEEPING WILLOWS

On Friday, November first, the day before the grand farewell, the Exposition opened without fanfare. Workmen hauled packed crates to railroad cars and concession owners tried to pretend the fair wasn't in the throes of ending. Exposition buildings and Midway acts opened and closed on schedule, and demonstrations and drills went forward. On the Grand Court at night, fountains sent up plumes of colored mist; on the edges of buildings, the luminous jewels of incandescent lights waxed and waned.

Alice Cenda wanted nothing more than an ordinary day, where hours passed without incident, and people, particularly the Bostock people, took little interest in her work. She got it. She dutifully performed as the Doll Lady in her salon, greeted and dismissed her audiences, ate her supper, and in the early evening settled into her apartment. A Bostock caretaker fastened her doors.

Several hours later, at 11:30 p.m., a carriage with three men pulled up on Elmwood Avenue behind the fence on the backside of her concession. Tony got out and entered the grounds at the West Amherst gate. He moved from shadow to shadow and threaded his way behind the Exposition Hospital and the scenic railway. Mingling with late-night cleaners and sweepers and passing by weary performers, he approached Chiquita's ticket booth. It was dark, but he was able to look through the window of the booth down to the floor.

"There she was," he recalled. Maneuvering her through the opening—no bigger than a cat door—wasn't as easy as they had hoped, however, and knowing that Bostock's men were on the prowl, they worked fast. Tony pulled and Alice struggled, and she wriggled out. Tony carried her to the fence and signaled to his friends. The two men—fellow workers in the Indian Congress—were there, waiting. "I boosted her over," Tony said, "and they caught her." The only casualty was Alice's dress, which snagged on the top of the barbed-wire fence. Once Tony made it over, the carriage pulled away.

Judge Thomas H. Rochford had probably seen a lot of hasty marriages in his time, and had helped tie the knot for dozens of seemingly mismatched people, but this might have topped them all. He had gone to bed when the knock came at his door. He could see a carriage outside, waiting, and he invited the couple inside. Tony said he was of legal age and Rochford knew Chiquita—he had been to the fair with his family. Tony didn't have money, but he promised to pay the judge later, and Rochford was willing to help. He interviewed the couple, signed a marriage certificate, and watched them drive away.

Maybe it was the commotion of escape that alerted the Animal King to Chiquita's disappearance. Within the hour, the showman knew she was gone. He sent for bodyguards and the Exposition police, and his lies slid out as smooth as butter. He had been startled awake, he said, by "cries of help, coming from Chiquita's headquarters." He had thrown on some clothes, raced to her rooms, and, much to his horror, seen "evidences of a struggle having taken place in her room." There was only one conclusion: Chiquita had been kidnapped. He needed the help of city police. Minutes later, a squad of Buffalo detectives was on the hunt.

The mistake the couple made was to go back to the Exposition.

They wanted some things to take on their honeymoon trip, Tony recalled. But when they pulled up to the West Amherst gate, they could see they were done. A Bostock clown, still in greasepaint, moved out of the dark and blocked their way. "He shoved a gun under my nose," Tony said, and, as the clown wrenched Alice away, he shouted words at Tony whose meaning we can only guess. "You to the weeping willows," he yelled, and ran off with his prey. Chiquita could do little to resist. "For once," she said, "I wished I was big."

Chiquita and Bostock.

When she was delivered to Frank Bostock, Chiquita cried, and apologized. That wasn't quite enough for the Animal King, though. He wanted to teach his Doll Lady a lesson. Being four feet taller than she was, and probably two hundred pounds heavier, this was not difficult. He knocked her down, and she fell to the floor, "senseless." Some of Bostock's employees wondered whether she would survive.[7]

VI
THE GRATEFUL GUESTS

Weather forecaster David Cuthbertson, when asked about prospects for Farewell Day at the Exposition, offered a mixed opinion. "There are indications of rain, possibly by tomorrow evening," he announced, "but there are reports that cause me to hope for a change." Cuthbertson's forecast was impossible to get wrong. November second dawned as a sunny day, and the only clouds that appeared were flighty and meaningless.

At the booths and shows on the Midway, the day began with good-byes. Workers who had labored side by side for more than six months offered good lucks for Charleston or St. Louis, exchanged souvenir picture books, and collected autographs. They produced more gifts for their managers—diamond rings, stickpins, and loving cups. And they grew nostalgic. A veteran spieler on the Midway could hardly hold back his tears. The guards, the ticket takers, the chair boys had all done so much, he said, and worked so very hard. "There ain't very many shows I hates to go way from . . . but this is the exception which proves the rule," he asserted. He hadn't expected it, either. "Buffalo people and all the people up in this cold, chilly region are half dead and don't know it," he declared. But "give them a little more time than anybody else to do a thing and you will git it done bettr'n anybody else can do it. . . ."[8]

As Midway performers departed, Buffalo leaders and the press beamed with satisfaction. The fair, they said, had made such a difference for some of these people. They had come to Buffalo untutored, poor-mannered, and unkempt, and, by the end of the season, they had been transformed.

The "Esquimaux," they said, had been "like so many children." But now many spoke English and almost every male had bought himself an American suit of clothes. (In fact, many of these Canadian performers had been on the American fair circuit before.) The press claimed that the Filipinos had also blossomed, and some distinguished local women had even taken them into their social circles for meetings and meals. The best part of all was that these newest "possessions" would take recently acquired habits back to their homes. When they returned to their islands, they would introduce civilized customs and convey "the benevolent purpose of the American Government."

Africans, too, would carry new habits across the Atlantic. "One great ambition of the native inhabitants of Darkest Africa on the North Midway is to be civilized," commented an *Express* reporter. There was, he went on, such a desire among the Africans to adopt Western manners and dress that quarrels among performers had sometimes grown violent. "The taunt that Chief So and So was more civilized than some other chief," he said, "brought out the knives and clubs."

Another newsman remarked that some African performers, having arrived in a crude and "primitive" condition, now walked the streets of Buffalo "clad as are American negroes, with a bearing that is stronger and better. . . ." He claimed that at least half of the people in the African village hoped to remain in the States. In fact, eight Africans had recently broken away from their guard and were seen boarding a streetcar before they were caught.

Some of the men and women of the Midway villages left with gratitude and envy of America to be sure. And some took away expertise

Darkest Africa performers with organizer, Xavier Pene.

in marketing or technology that they could employ to their advantage at home. Yet many probably had mixed feelings. They had found out what it was like to be dark-skinned strangers in a northern American city. They had been gawked at, scoffed at, and scrutinized. They had spent months performing for people like Mabel Barnes—earnest, well-meaning white people—who had studied them as specimens of subhumans.

It wasn't just Africans or Filipinos or Indians who found themselves on the opposite side of the looking glass or on the receiving end of stares. Ben Ellington and other Southern performers in the Old Plantation were certainly aware of how much the progressive black community in Buffalo disliked what they did to make money. They were also alert to the way that Northern people, of all backgrounds, dismissed them. Ben reported that his tips from Northern customers were far inferior to those offered by Southerners. He was eager to get back home to Georgia.

Then there were performers who would have been happy to take American "civilization" and to eliminate it altogether. The infamous Filipino Pablo Arcusa had brought to the fair little liking for American beliefs and policies, and another insurgent, Gregorea Tongana, had been discovered in Rainbow City in the fall. A banduris player who deeply resented American occupation of his homeland, Tongana had served on the staff of General Malabar of the Philippine resistance, and he later auditioned in Manila for a part in the band. Other band members knew who he was, and, until he fled, gave him help and cover.

Finally, there were those who never made it home. Historians have explained that this wasn't just bad luck, but rather a side effect of imperial encounters. Along with guns and money, Westerners had for years carried to indigenous peoples their invisible baggage of germs. In this case, the routes were reversed. In Buffalo, tuberculosis swept through the Indian Congress and sickened eight people. In the Filipino Village, it killed Tastuala Ruyes, twenty-five years old. It also took the life of Henak from Labrador. Mumps infected the Hawaiians and measles hit the Esquimaux Village, striking eleven Natives and killing seven-month-old Sebelia Nikolenik. Others became infected with diseases en route to the Pan-American fair. Social Darwinists at the time said that events like these gave a firsthand demonstration of the survival of the fittest and showed why some groups were vanishing. To them, every death was a lesson in evolution.[9]

VII
FAREWELL

Mabel Barnes had first clicked through the turnstiles at the Pan-American gates at the very end of April. The grounds had been muddy then, the piles of lumber hard to navigate. She found exhibits closed, gardens bare, and pedestals still waiting for statues. Now, thirty-three visits later, on a cool, sharp-edged November day, she had come full circle. The city she had loved was on the verge of disappearing. She wandered by boxes, barrels, and overturned dirt. One building, the stately New England exhibit, had been gutted by a fire, and workmen with crowbars and axes wrenched apart its porch. Others, like the sixteen buildings in Bostock's show, had been auctioned off two days earlier. Esau's quarters and Chiquita's living and reception rooms, along with animal skins—from lions, tigers, panthers, leopards, and jaguars—had been sold.

The schoolteacher entered the fair alone that final day, joining 125,000 others on the grounds. She walked to the New York State Building, where, four months earlier, she had waited on tables at a reception for fellow teachers. She took a last look at the Esplanade and the Court of Fountains. Abby Hale then met her, and together they said good-bye to the big exhibition buildings. As usual, the women saved the Midway for last. They wanted to see one final show—the one that had enthralled them more than any other: Darkest Africa. But it was too late. The stockade surrounded a deserted encampment. "To our disappointment," Mabel said, "we found [it] empty and forlorn."

They went back to the Electric Tower and "feasted [their] eyes on the beauty of it." They tried more buildings and found more of them locked.

As the two women toured the grounds that afternoon, they would have noticed crowds clustering around special visitors. Carrie Nation was said to have been there. Audiences loved showing how much they hated Mrs. Nation, with her loud opinions and saloon-smashing attacks against the devil, alcohol. The six-foot-tall temperance reformer, who spent much of her outspoken life in handcuffs, would have met plenty of enemies at the Pan-American. Two months earlier, while lecturing at Coney Island, Nation had fixed her fierce opinions on President McKinley, who at the time lay injured in Buffalo. She said publicly that she was "glad" he had been assaulted. "The President was a friend of the rumsellers and the brewers," she declared, "and therefore did not deserve to live." When the New York audience hissed at her, she told them they were "hell hounds and sots."[10]

Abby Hale and Mabel Barnes had no time for the likes of Mrs. Nation. They also skirted another celebrity. Holding court on the West Esplanade bandstand, with her big oak barrel, a black cat, and Captain Billy Johnson, was Annie Taylor.

All her work over the previous weeks—the attention, money anxieties, and most of all the most frightening ride of her life—had brought Taylor to this triumphant moment. Resplendent in a blue jacket, and showcased next to Iagara, the cat that allegedly had preceded her over the falls, Taylor began greeting people around noon. She took particular time with children and those "who displayed enthusiasm above the average." She also tried to make it clear she was not putting on an act at the Exposition, like Midway performers. She was, for the sake of $200, "receiving" visitors.

And what a reception it was. Around and behind her, a line snaked with curious and admiring (and paying) guests. Thousands of people of all sorts—"a continuous stream of humanity"—had come to meet her. Exposition guards, reported a witness, "had their hands full preventing crushes."

Observers said she showed signs of her ordeal. Her nose was red, and she looked ill. She didn't want to shake hands—her doctor, she said, advised against it. And she admitted that she ached. "I am very much bruised as the result of my trip," she said to a reporter. "The muscles of my back, between the shoulders, are particularly sore." She couldn't stand the cold, either, and had to stop and warm herself now and then inside the Temple of Music. All in all, one reporter commented, Mrs. Taylor seemed rather the worse for wear. He also thought she had not aged particularly well. "Mrs. Taylor looks fully the 43 years that she admits," he commented.

Annie Taylor had won her first battle—she had subjugated Niagara Falls just as surely as the rich men who had diverted its water for modern technology. But she was struggling in her second battle—her fight for dignity. She had received plenty of compliments, to be sure. "All who have met and talked with her agree upon two points," said the influential *Buffalo Express*. "She is certainly the best-educated and most refined appearing person who has ever attempted a feat of daring at Niagara. And she is (or was before her violent experience) the calmest, most self-possessed of all the great multitude of adventurers commonly classed as Niagara Falls cranks." The *Express* also described her as "modest and sensitive." Other papers called her the bravest woman in the world and suggested that she had put to shame other male Niagara daredevils, like Carlisle Graham. One columnist also pointed out that, given as she was driven to desperation, she chose a more honorable route than suicide. She had simply sought a "competency for the remainder of her days."

Still, there were plenty of people who maintained that Mrs. Taylor's barrel ride revealed nothing more than dumb luck and cheap showmanship. "The net results of the hazardous trip," wrote the *Grand Rapids Herald*, "is that Mrs. Taylor has acquired a little notoriety, which she can capitalize to the extent of becoming a dime store freak." Other newspapers called her "reckless and selfish" and ridiculed her audiences. "The country is filled with people who are just morbid enough to crave a look at her," remarked one paper.

As the general public debated Taylor's character and judgment, the distinguished women of Buffalo seemed to ignore the adventurer. They spent the final days of the fair in the elegant confines of the

Women's Building, enjoying last luncheons, toasting each other with crystal, and nodding and smiling over bouquets of pink cosmos.[11]

As dusk approached on Farewell Day, Mabel Barnes and Abby Hale ate a picnic and walked together toward the Triumphal Bridge. The November sky, cold and luminous, performed its own send-off at sunset, and then, at a little after 6 p.m., the Illumination took over. Every band on the grounds struck up "The Star-Spangled Banner."

The women could not get enough of the scene. "We looked and looked and looked," said Mabel. The schoolteacher rarely repeated herself, so it must have been a powerful moment. They wanted to stay until midnight, when John Milburn would throw his extinguishing switch. But Mabel knew what the end of a fair usually brought with it, and already she could see signs of destruction, as rambunctious visitors began to rip up flower beds. Not willing to get caught in the mayhem, and perhaps not willing to see the ruination of everything she had admired, Mabel went with Abby through the Midway one final time. Then they left.

The women knew what they were doing. Even before they walked out, scavengers began raiding buildings for food and tore away signs and banners and decorations. In the Illinois display, said one witness, crowds swept the guard "aside like a fly" and men held him down while they stripped the apple exhibit of every piece of fruit. In the chrysanthemum exhibit, "souvenir fiends" took every flower.

For those who could concentrate on it, the fireworks show proved stupendous. At 8 p.m., on Park Lake, the reliable James Pain sent up a sky-full of meteors and let loose a shower of exploding stars, rockets, and shells. He lit up a portrait of John Milburn and re-created the bombardment of San Juan. At the very last, he sent up a poignant set of illuminated words: FAREWELL TO THE BEAUTIFUL CITY OF LIGHT.[12]

The final tribute to the fair, delivered at 11:30 p.m. in the rotunda of the Temple of Music, was, according to at least one listener, "a funeral oration." The mourners, huddled against the evening cold, included Latin American representatives, state commissioners, and all the fair's backers and planners.

Mayor Conrad Diehl spoke first. He accepted the gift of the Temple of Music organ—one of the few valuable souvenirs the city would retain—from Buffalo's James Noble Adam. He then turned over the podium to the man, once his political adversary, who had since become his trouble-tested colleague: President John Milburn. "We started out," Milburn said to the group, to "begin the 20th century by a binding of the relations of the free peoples of North, Central, and South America. . . ." This, he said breezily, had been accomplished, and he thanked the Latin American emissaries. And this land, he went on, gesturing at the grounds, "was once a plain and a scheme." But it had become "one of the most beautiful scenes that the human eye ever has looked upon." He thanked the people who built the fair.

Then he entered rockier terrain. "The exposition," he announced, "has been a great success." Was there a collective intake of breath? Milburn knew insolvency was on everyone's mind, but he refused to give in. He would not use the word *failure*. "If it has been a financial unsuccess," he said, "it has been worthy of all praise." He tried it again. ". . . I believe that what financial unsuccess there is about it the citizens of Buffalo will meet with the same energy and generous spirit by which they constructed the exposition." At the very least, he asserted, "we have made Buffalo known to the United States as it never was known before."

Milburn finished speaking just before midnight, and, with a smaller group of men—including Director of Works Newcomb Carlton and Henry Rustin, mastermind of electricity—he walked to the Triumphal Bridge. High up in the Electric Tower, eight army buglers rang out "Taps." Except for some muffled weeping, the grounds were still. Milburn touched a button on a box. Connected to the Exposition's rheostat, it sapped the power from the fair's 160,000 incandescent bulbs. Gold to crimson they went, and then shell pink to pinpoints. The Exposition became so dark, said one witness, it seemed like stars had "fallen into a sea of ink."

The end, a reporter wrote, was "apparently painless." It is hard to know why he used personification to describe the final moments of the Exposition, but, then again, death and electricity had been so tightly linked that autumn it was perhaps unavoidable. Rainbow City perished, he said, "like a fair queen, a lovely lady, from whose cheeks the color faded, in whose eyes the luster died, on whose lips the bright smile vanished. The darkness of night enveloped her like a funeral pall. She did not struggle, no sigh, no moan. The fountains of her life fell and grew fainter, until they stopped."

To the less overwrought, it simply became very, very dark. Fair-goers made their way to the exits, some of them stopping to remove tiny lights and take them as souvenirs. On the Midway, other visitors began to act out. They smashed booths, threw restaurant chairs, and fought each other with vines and branches, plants and palms. They threw confetti until they waded in pools of paper. They took food and fruit, ripped feathers from Indians, and wrested flowers from Hawaiians. Twenty-five of them became the last patients at the Exposition Hospital.[13]

While crowds surged through the Midway, behind the doors at Bostock's Wild Animal Arena, Chiquita prepared for the night. She had entertained audiences all day long. If she carried any bruises on her body, they were concealed, and if she was distraught, she had put on a brave face. She had moved through her drawing-room act as though nothing had happened. It wasn't as though she had any choice. "I was taken possession of," she said.

For Frank Bostock, the outlook was bright. Having spent months grooming the press, he was able to shape the story of Chiquita's flight to his liking. Tony Woeckener, he explained, had facilitated the abduction of Chiquita, and he had been arrested. A competitor on the Midway had sought to undermine the Animal King, and Tony had been the "tool" the schemers used. Thankfully, a loyal employee had intervened and rescued the little performer. So happy was he to have order restored, he decided not to bring charges against the "kidnappers."

As for Chiquita's affection for the cornet player? "Chiquita has had these little attachments frequently," he told reporters. "Then when the suitor thinks he's the whole thing, her love grows cold." And the wedding? It had been forced. The Woeckener boy and his friends had taken Chiquita to dinner and had fed her some wine before driving her to the judge. Even she would admit she was in a "batty state." Chiquita was worth a lot of money. She had thousands of dollars and jewels. Who wouldn't want to marry her? The Animal King announced he would do what he could to protect her from these sorts of manipulators in the future.

That protection, in fact, would begin right away. Aware that Tony had begun to protest, the showman would put Chiquita out of his reach. Luckily for Bostock, this was Buffalo. Canada was only a river away.[14]

10.

The Elephant

I

THE WARRIORS

"In all that followed," remembered Chiquita later, "I did not lose my courage."

Frank Bostock, with the help of his wife, smuggled Chiquita through Niagara Falls and across the Canadian border. This made things so much easier for him in court. When he was sued by Tony and his father to relinquish his star performer, he claimed he could do nothing. He didn't have Chiquita and he couldn't bring her to Buffalo. The judge was convinced. On November 8, he decided that Alice Cenda was beyond his jurisdiction.

The judge might also have been swayed by reports of the showman's generosity. Bostock's lawyer told him that the Doll Lady was "never kept in restraint," and that she had been whisked away by Bostock's nurturing family in order "to recuperate." The lawyer asserted, too, that Bostock "treated [Chiquita] as a member of his family, that he had bought her fine raiment, jewelry, horses, and carriages and all manner of luxuries." He had even taken pains to educate her. Wasn't it apparent that Bostock was the one being badly used? Wasn't it obvious that there was a "conspiracy" afoot to steal away his moneymaking show?

Tony's lawyer had a firm response. Chiquita never owned fancy clothes or cars. They were on loan from Bostock. As for the claim that the animal trainer truly cared for the performer? Bostock, he said, "has cared for her just about the same way he has cared for his elephants."[1]

There had been nothing but positive reports about Bostock and his elephants all season. The showman had boasted nonstop of their tricks and performances and the press had reported nothing about mishandling or abuse. Tony's lawyer, it seems, was referring to very recent news.

On November 6, Bostock had sent the indefatigable Captain Maitland into the offices of the *Buffalo Enquirer* to make an announcement. As usual, the writers readied themselves for a barrage of exaggeration. This story, though, seemed different. In fact, the newspaper staff suspected that the captain might be speaking the truth. Maitland told them that Jumbo II, Bostock's enormous elephant, was going to be killed in public. The veteran animal–soldier, the beast decorated for bravery in the Ethiopian theater, had developed musth. It seizes some elephants, Maitland explained, and if action were not taken, Jumbo might "wreck the entire exposition." Furthermore, he said, "people are in danger of their lives and we do not know how to handle or control him. Therefore, he must be killed."

Maitland did not need to remind the writers of very recent events—they were well publicized. A few days earlier, just as the Animal King was beginning to dismantle his show, Jumbo II had acted up. Henry Mullen, the assistant elephant keeper, had entered the animal's cage to clean his bedding when Jumbo wrapped his trunk around Mullen, picked him up, then dropped him. Then he did it again. The second time, Jumbo twirled Mullen "about as a juggler spins a plate," and tried to step on him. Mullen yelled, and other assistants, armed with "pronged spears and clubs," beat the elephant until he desisted. Mullen suffered broken bones and lacerations. It would have been worse, the keeper said, but the elephant, being heavily chained, couldn't use his tusks or lift his feet high.

And there was more.

On November 6, twelve-year-old Tina Caswell, the daughter of Bostock's zoo manager, decided to visit Jumbo and feed him some hay. For some reason, Tina had "forgotten the elephant's bad nature." Jumbo took the girl in his trunk and hoisted her. When her dress tore, she fell to the ground. She was badly frightened.[2]

Within a week, Jumbo II went from being a magnificent beast to a "murderer."

For months, he had been the wonderful warrior elephant, the pachyderm with a history of saving men and inspiring his own species. True, he had been rambunctious. Even before Buffalo he had strenuously objected to being placed in slings and hoisted by derricks onto ships, and he had never liked being confined in crates. He had also been overly boisterous in wooing Big Liz, the female elephant in Bostock's zoo. On one occasion in July, he tried to break the chains that bound him head to foot, and he stomped on a tub of liquor meant to subdue him.

But, the press explained at the time, his unruliness had simply been the product of "cupid's dart." His behaviors had simply been "antics." And ever since then, he had quietly entertained thousands in Rainbow City. Big pussycat that he was, he had been unnerved by thunderstorms and had even trumpeted in fright at the sight of a tiny marmoset. But all that was gone. Now he was the Exposition's monster.

While some officials agreed that Jumbo II was afflicted with musth—which male elephants periodically experienced when ready to mate—others declared that he had been wicked all along. The famous war record that had listed his heroic and altruistic deeds was now understood as a "very bad record." A new report asserted that he had killed seven men in India. Outside of Buffalo, the fabrications of the press went further, announcing that he had killed seven people at the Exposition. Indeed, there were now hardly enough terrible words in the English lexicon for Jumbo II. He was a "huge brute," a "rogue," and, above all, a "man-killer."[3]

Nobody asked why young Tina Caswell, who had been in the menagerie for months, was permitted to wander into Jumbo's pen. No one challenged Mullen's story. And no one wondered why Jumbo was sentenced to death when other ferocious animals were permitted to live—in fact, their aggression was a drawing card. People flocked to see Rajah, the "man-killing" tiger. In April 1901, the Bengal tiger had leaped upon Frank Bostock in Indianapolis after training exercises, ripped the showman's arms and legs, lacerated his face, and thrown him, unconscious, on the ground. Rajah, who had killed a young trainer just a month earlier, was nevertheless forgiven, and soon was dispatched to perform in Buffalo.

Then there was Madame La Belle Zelica and a new lion named Trilby. In the beginning of October in Rainbow City, Madame Zelica entered a cage with four lions and ordered them onto pedestals. Trilby complied—but became "sulky." When the trainer's back was turned, Trilby leaped onto her, dragged her to the floor, and slashed her face and shoulder. Come Buffalo Day, though, Madame Zelica, her arm in a sling and her face bandaged, and the lion, not necessarily contrite, were featured in a new act. The trainer's brave reappearance, it was announced, "would be the most sensational performance of the year."

Jumbo II, however, was a different case. Maybe it was true that elephants who went into musth were too hard to handle. Or maybe this big elephant was close to the end of his run at Rainbow City and was expensive to box up, haul to the train station, and ship to a different state.

Captain Maitland ensured that Jumbo II's demise, or the means of it, was the talk of the town. Newsmen speculated about how the big animal would be killed. Some bad elephants, they knew, had been poisoned. Tip, an elephant at the Central Park Zoo in New York City, had been fed cyanide. Sport, on the other hand, one of Bostock's elephants who had been injured, had been hanged by a steam derrick in Baltimore. Hanging, though, had its problems. The editors at the *Courier* recalled one sad scene when zoo elephants were forced to pull the rope that asphyxiated their companion. Then there was shooting. Maitland himself suggested this possibility, but he despaired of finding an elephant gun.

Finally, there was electrocution. Niagara's power was so close and plentiful.

How much Bostock took inspiration from the public's fascination with Leon Czolgosz's end at the Auburn prison is unclear, but he couldn't have missed seeing how the event took hold of people. And he was certainly aware of how, to some, this massive shock would be a fitting end to the fair. An Indian beast brought down by a British animal trainer with the help of the most modern Western technology? To many people, it would seem a perfect conclusion.

Bostock set the date for November 9 at 2:30 p.m., and, soon enough, Captain Maitland broadcast the occasion. The show would take place in the Pan-American stadium, and special trains would run to the site.

Notices also explained that admission would be charged. Bostock was asking fifty cents per person. Visitors needn't worry about squandering their hard-earned cash, though. It would "be the sight of a lifetime."[4]

A reporter, referring to the public's love of seeing others suffer, remarked that Bostock had conceived of a "spectacle fit for a Roman holiday." Another man, a press agent, didn't want his comments put into print. When asked whether Jumbo was really to be killed, he said he didn't know. Then he added, "I think it's a shame to kill him . . . but I don't want you to publish it." Of course he didn't want them to publish it. The speaker was Jack Maitland himself, the Animal King's greatest asset.

Captain Maitland knew that there would be plenty of people eager to see an elephant brought to its knees. Thousands had paid to see the killing of local canines at the Indian dog feast, and this would be an even bigger marvel. There was something exotic about it, too. Killing an elephant in a stadium wasn't exactly tracking an elephant in the bush, but it was as close as many would come to this elite and fashionable sport. Accounts of big-game hunting in the American West, South Asia, and Africa had a wide audience at the turn of the century, and details of stalking buffalo, bears, lions, tigers, and elephants reinforced prevailing ideas about civilization and empire.

At the same time, many Americans came to know elephants not as quarry but as amiable and clownish animals. A ship's officer named Nathaniel Hathorne, the father of the famous writer, was reportedly the first to bring a living elephant to the United States, in 1795. Sailing from Calcutta, he lettered the word "ELEPHANT" in the logbook, capitalizing his excitement. The crew took the two-year-old female ashore in New York and she proved to be "a great curiosity," touring up and down the eastern seaboard, hoisted in and out of schooners.

Word has it that a New York entrepreneur, Hachaliah Bailey, encouraged by this exhibit, bought a second elephant, "Old Bet," and put her in a menagerie. Old Bet, it seems, didn't live long, but accounts of her demise—she was apparently shot—vary. What is certain is that Hachaliah Bailey put up a monument to her that still stands, in Somers, New York, in front of his Elephant Hotel. And what is also clear is that Bailey's animal show inspired a young Phineas T. Barnum to start a menagerie of his own later on.

Elephants were such crowd-pleasers that, before long, traveling shows and circuses could not be without one. Affection for elephants echoed—among the middle class, at least—a rising sympathy for animals in general, as well as a sense that elephants shared with people a capacity for intelligence and empathy. Old Bet herself was pronounced to be nearly human. When her owner returned after a ten-week absence, the animal cried out and caressed him with her trunk. Bet was so obedient that she traveled without ropes and chains and drank "all kinds of spirituous liquors; some days . . . 30 bottles of porter, drawing the corks with [her] trunk."

The American public's infatuation with elephants reached a crescendo in the late 1880s, when people met and fell in love with Jumbo. This pachyderm, who gave the English language a synonym for "enormous," was captured in East Africa in 1861 and later raised and trained at Regent's Park Zoo in London. Jumbo was known for his (usually) sweet temper, his love of peanuts and buns, and his willingness to carry children on his back. Queen Victoria's children and grandchildren rode Jumbo. Young Teddy Roosevelt rode Jumbo.

In 1882, though, much to the consternation of English children, the seven-ton, eleven-and-a-half-foot-tall elephant was sold to P. T. Barnum and shipped to the United States. Barnum, who provided Jumbo with American gingerbread instead of British buns, had bought himself a moneymaker. People loved Jumbo, and they were grief-stricken when, in 1885, in Ontario, Jumbo was killed. Reports of the way that a locomotive accidentally slammed into the twenty-four-year-old elephant heightened their misery. Rumor had it that as the freight train bore down upon him, Jumbo moved a smaller elephant out of the way. Just as wrenching was the story of the dying Jumbo wrapping his trunk around his trainer.

Bostock tapped into the fondness for elephants at his Pan-American animal show. He ordered Big Liz and a set of younger elephants to carry visitors around the Midway almost every day, and to perform tricks and engage in races. He showed off baby elephants who played seesaw, rolled barrels, and walked on their hind legs. And, for a long time, the showman sought public affection for Jumbo II, the "handsome" warrior. But not anymore.[5]

II

MRS. LORD'S ARMY

On November 3, a reporter for an Erie, Pennsylvania, morning paper, eager to learn what had transpired in Bostock's other newsworthy battle, rapped on the door of Tony Woeckener's house on East Sixth Street. Ernest Woeckener, Tony's father, a laborer at a local iron works, welcomed the newsman inside, where they navigated through a maze of instruments—a bass viol, a cello, a guitar, a banjo, two cornets, a tuba, and three violins. The older Woeckener, leader of the family band of seven musicians, ranging in age from five to eighteen, regaled the visitor with a short concert. With instrumental backup, a ten-year-old Woeckener girl warbled "Where the Apple Blossoms Bloom." A five-year-old followed with "Come and Keep House with Me." The Woeckeners could offer no news about Chiquita and Tony, however, and the reporter left empty-handed.

In Buffalo, on November 8, another reporter had better luck. He found Tony in a hotel downtown. Without prompting, Tony spoke of his desperation and pulled out a letter from Chiquita. Written in tiny script, it began with "My Dearest Sweetheart" and concluded with a cluster of crosses, row after row. "These are kisses," Tony explained. "Let anyone who doesn't think she loves me look at that. And I've got a whole lot more just like that one in my pocket." As for giving up and letting his wife go? He wouldn't. "If it costs my life," he said, "I'll get her." He had to admit, though, that Bostock was outspending him by far, and his family was telling him it was time to come home.[6]

By early afternoon on November 9, under heavy, cold skies, more than a thousand visitors made their way onto the fairgrounds. The Exposition was mostly a façade now—a stage set of empty pavilions and a Midway littered and muddied. At least the Electric Tower stood intact, a monumental reminder of the grand season. By two o'clock, the numbers of spectators had grown, as streetcars disgorged more and more people eager to see the end of the big elephant.

In the stadium, electricians hooked up power. The public was probably familiar with the language of electrocution by now, but the press, always ready to share sensational details, offered reminders. Step-up transformers would deliver alternating current of eighteen hundred to eleven thousand volts of pressure to the elephant. Jumbo II, bound with chains and tied to wooden posts, would wear a harness that allowed electrodes and moistened sponges to fit snugly into his ears and mouth and near his tail. This, officials explained, was where the animal's hide was most thin. They also said that they did not expect the process to take more than fifteen minutes, and that the end would be determined, just as it had been at Auburn, when physicians ascertained that "life was extinct."

Even as the technicians worked, though, people outside the grounds grew more disturbed. What had started as a murmur of dismay began to swell into anger. City residents, by the dozens, picked up phones and called Mayor Diehl and Exposition directors. They sent the Humane Society 150 messages. It wasn't so much the killing of Jumbo, these protesters said, but the *public* killing. The head of the local SPCA, DeWitt Clinton, agreed. The humane killing of the elephant was "none of our concern." But making it a show? "It was an outrage on public decency."[7]

———

In taking on Buffalo's SPCA, Frank Bostock faced an opponent as fierce and determined as any he might ever encounter. The second-oldest animal-welfare society in the country, it had put together a record of tough-minded intervention.

It all began in the late 1860s, with a small powerhouse of a woman named Mary Lord. Appalled by seeing thin and beaten-up mules pulling barges along the Erie Canal, Lord wrote to Henry Bergh, founder of the country's first ASPCA in New York City, to describe what she had witnessed. Bergh was incensed. "The barbarity exhibited in the treatment of these unoffending animals," he wrote back, "is a deep stain upon our boasted civilization. Something should, something must be done." Soon enough, Lord had an organization and helped charge the bargemen with cruelty. To make its case to a judge, Lord's group brought a stand of mules directly to the courthouse and pointed at the "galls on their necks" and their "starved and emaciated bodies." The judge delivered a quick verdict of guilty.

A fireball of "boundless sympathy," Mary Lord, accompanied by her collie, trotted through city streets in a small phaeton drawn by four Shetland ponies, sniffing out mistreatment. She and her colleagues stood at stockyards to make sure that cattle, no matter their intended end, were transported with care and treated with decency. They opened a pound and visited every school in the city to talk about the needs of "dumb animals." The society had formidable backers. One, an elderly Buffalonian, explained that he had taken up the cause of animal welfare when he was a young boy. He had killed a mother bird, and his father pointed out to him that the babies would now all perish. He never forgot that, and from then on never killed a living creature. He was one of the city's most famous residents: former US president Millard Fillmore.[8]

Mary Lord died in 1885, but the SPCA, despite being poorly funded, continued its work and in 1901 took on the challenge of the

Mrs. John C. Lord.

Mary Lord, friend to animals, honored in stained glass.

Pan-American Exposition. Its agents saw some problems right away. Horses pulling carts and wagons to the building sites were overworked and overheated. To the amusement of the public, the society equipped the animals with straw hats fitted with moist sponges in the crown.

Throughout the Exposition season, society members fanned out over the grounds, inspecting animal exhibits. They targeted the Midway's Streets of Mexico because of its bullfight, but they soon realized that the bulls were exceedingly tame and there wasn't much of a fight. Their efforts, though, caught the attention of a local editor, who reminded his readers that at the Pan-American, whose purpose was to show human progress, there was no place for "torture, either of animals or human beings."

The Humane Society discovered more serious issues elsewhere on the fairgrounds. The bears in the Idaho exhibit were covered with flies, and the ponies in the Indian Congress were sore and thin. Society members demanded that the ponies get more food and better oats, and, weeks later, were gratified to find them fat and well. The reformers also cited the Beautiful Orient concession for beating donkeys and camels and for forcing camels to kneel on hard stones to take on riders. They ordered cushions.

They did not stop there. They pointed out the cruelty that provoked elks into diving headfirst into small tubs of water. An SPCA agent visiting the Midway show discovered that the animals' diving was hardly voluntary. A "sharp instrument" hidden in the trainer's hand forced the elks to make the long leap into tubs. The show's manager was warned against repeating the offense.

The society also identified abuse directed at Bonner, the mathematical horse. After someone heard cries coming from the horse's stall in August, the group discovered that Bonner's trainer whipped him into performing sums and other calculations. The trainer apologized, saying he hadn't realized the horse's mouth was bleeding.

For all their attentiveness, it wasn't until September, when their agent spotted a bleeding lion, that society members focused on Frank Bostock. They had missed the experiment of Jumbo II and Tiny Mite. They had also missed seeing the showman send his crocodile Ptolemy over Niagara Falls.[9]

Maybe the society would have paid more attention to the Animal King if he hadn't been a charming Englishman. British and American anticruelty societies focused on rescuing animals but tended to target perpetrators in immigrant and working-class communities. Formed by the urban well-to-do, the societies were sometimes as devoted to human social reform as to saving animals. In Rainbow City, the Humane Society followed this pattern, expending considerable energy on infractions in the Indian Congress and the Mexican and Orient concessions. It criticized the dog feast not because it was cruel to canines but because it was an "unedifying" spectacle. As for the doings of the English Animal King? The witty Captain Maitland showered the press with flattering copy that spoke of Bostock's sophisticated pedigree. For the most part, Bostock slipped by.

Perhaps Buffalo's animal-protection officers would have paid more attention to the Bostock zoo if they knew that even in "civilized" England, the menagerist had been taken to court for animal abuse. In 1890, three years before he sailed for the United States, Bostock was charged with cruelty to eight horses. In fact, out of the forty horses he used to haul his menagerie from town to town, two-thirds were said to be suffering from open wounds. One had a cut thirteen inches long. Bostock sent a letter to the Leyland police court, admitting his guilt, but defended himself by saying that an inspector in Blackpool had told him his horses were in "workable condition." Unfortunately for Bostock, the Blackpool inspector showed up in court, too, and declared that Bostock was telling a barefaced lie. Other witnesses reported that Bostock was more willing to pay the cheaper fines than to give his horses better treatment. The judge gave the showman the maximum penalty and ordered him to pay the fees of a veterinary surgeon.

Although on occasion Bostock was arrested for cruelty to his show animals—such as the time he was accused of baiting and fighting two lions—it was the horses that hauled his equipment and cages, and were slaughtered for food, that were most commonly injured. In the spring of 1891, Bostock ran afoul of the RSPCA for dragging an injured horse behind one of his wagons, and again he was convicted of cruelty. Seven months later, in 1892, the police sent him back to court for working horses that had open wounds. After another few months, a veterinarian testified that one of the showman's horses suffered from a separated biceps muscle. And so it went, Bostock leaving a trail of fines, a wake of hurt animals.

Two years later, he was in America.[10]

Inside his stall that November morning, Jumbo II ate hay. A few hours later, and over at the stadium, several thousand spectators took their seats on wooden benches. They could see killing equipment at the end of the field: a wooden platform, stakes, and big copper wires running to a rubber-handled switch on the stadium wall.

Then, suddenly, the crowd was told the show was off.

"Society people" had brought their influence to bear on Mayor Diehl and demanded that he abort the event. The city, they said, needed to be "spared a reflection upon its dignity." The mayor, himself an animal advocate, called Bostock to stop his show.

The spectators, some of whom had come from miles away, were angry. A few took advantage of the offered refund and left for home. Other people—maybe five hundred of them—refused to move. They didn't believe the execution had been canceled. They noticed that none of the electricians had left the platform. Then a rumor circulated that Bostock still intended to kill the elephant—it just wouldn't be a public event.

As the light faded and the weather cooled, the people stayed put. Then, at around five o'clock, they heard the sound they hoped for— the clinking of metal. Looking toward the entrance, they could see the big elephant being steered into the stadium. Two young elephants, themselves Exposition celebrities, strolled on either side of him. They looked, said one observer, like "a battleship convoyed by two torpedo boats." Jumbo II, encircled by heavy chains, placidly reached out to his small companions, and, pulling up grass with his trunk, offered it to the one named Roger.

Things moved quickly. Attendants wrapped ropes around the big elephant, positioned the electrodes, and tied his legs to stakes. They moved the little ones out of the way. Bostock spoke to the crowd and told them Jumbo was tearful. Jumbo flapped his ears. Someone reported that "streetcars on several lines had been stopped" to supply the voltage. The trainer raised his arms and the electrician threw the switch.

What the audience expected, it seems, was a spark, or smoke, or at the very least the hissing of steam. They counted on seeing Jumbo react, perhaps rear up, and then, certainly, they expected to see him sink in a thunderous heap onto the ground. But they saw nothing. The electrician opened the next switch. Jumbo apparently needed higher voltage.

The elephant flapped his ears, picked up a loose plank on the platform, and tossed it.

The electrician threw open both switches.

Jumbo swung his trunk.

In a split second, the mood of the crowd shifted. Spectators who had been eager to see a dead elephant suddenly began to cheer the live one. They laughed at the electricians and made jokes.

Jumbo remained standing.

After several minutes, it was clear the show was over. Bostock's attendants brought back the smaller elephants, reattached them to Jumbo, and walked the group down the concourse to the Bostock concession. Close behind them, the crowd followed, jeering.

Some witnesses declared it a sham. Not only had the switchboard never lit up, but the wires looked dead. There were no marks on the animal. Others thought it was an experiment that failed—the elephant's skin had been a nonconductor.

When reporters caught up with Frank Bostock, they found the Animal King willing to talk. Maybe he would try to destroy the elephant again, he said. Or perhaps he would see whether Jumbo could become peaceable. "He only gets wild in spasms," the Animal King admitted. "I will shut him up in a box and ship him in that fashion."

And he did. Led into a special car, and encumbered with chain, Jumbo II managed some "frolicsome antics" while in transit to Bostock's zoo in Boston. Once he rejoined his companions, Big Liz and Little Doc, though, he quieted and allowed himself to be paraded past five thousand spectators to the Cyclorama Building on Tremont Street. There he stayed through the winter, calm and steady.[11]

III
MRS. BADGER'S MISTAKE

If Frank Bostock suffered a public-relations disaster with his failure to do away with Jumbo II (assuming he did not stage the entire affair), he was winning in his battle with the world's littlest woman. He had hidden her. In mid-November 1901, a reporter in Erie, Pennsylvania—likely the same newsman who had been serenaded by the Woeckener girls—wrote a story about Tony Woeckener's missing wife. The headline asked, WHERE IS CHIQUITA? and the reporter issued a plea to the public to send Tony any information on her location. He also posted a sketch of Tony, looking wistful.

Tony did find Chiquita—not at the Charleston Exposition, where he thought Bostock would take her, but in Boston, being advertised for a show. Within days, he and his father were in New England, where Tony was allowed to see his wife only in short, heavily guarded visits. Meanwhile, having sued the menagerist again, the Woeckeners waited for a decision from the courts. On January 6, 1902, Tony's father sent a terse message back to Erie: "Hard fight on."

An associate justice of the Massachusetts Supreme Court listened to Tony make his case against Bostock, and listened to Chiquita. Alice told him that she did not like to break her contract with the Animal

King but wanted to be with her husband. The judge made a decision. The marriage between Tony and Chiquita, he said, "should not have taken place under any circumstances." The arrangement was not "the offspring of love or affection." There was "some ulterior motive." As for the little person herself? "The wife," he declared, "is not an ordinary person nor of ordinary capacity and intelligence for a woman of her years." He did insist, though, that she was perfectly capable of understanding—and fulfilling—her detailed employment contract. In sum, she belonged to Bostock.

Under oath later on, Chiquita recounted that, after Bostock had locked her in a Boston apartment, he had produced a new contract. If she did not sign it, she was told, "she would never be permitted to see her husband again." Faced with Bostock, his lawyers, and his employees, along with their "threats, intimidations, and curses," she signed the paper. The word that came to her mind when she recalled that moment was "terror."[12]

In the contract, Chiquita agreed to work for the Animal King for eighty dollars per month, through the close of the Louisiana Purchase Exposition in St. Louis in 1904. She could work for no one else. She had to accompany Bostock or his "agent" from "place to place, and work and exhibit herself as a midget." She had to "continue to live as one of [Bostock's] family" and to "obey all the rules and orders." Unlike other contracts that her father had helped negotiate, and that had been written in both English and Spanish, this one was produced in English only. She was not given an opportunity to look it over.

Bostock did agree to offer Tony a job with his zoo. But Tony couldn't bring himself to take it. "I thought he wouldn't hesitate to send me on some duty connected with his animals," the musician said, "which would incense them and they would tear me in pieces." He was serious. His father had seen enough of the Animal King to believe the same thing: "We knew too much to be drawn into such an agreement. Bostock would have killed Tony in six months."

Tony left Boston to go back to his family in Erie, and, not long afterward, Madame Morelli sent word that she and Alice were boarding a ship in New York Harbor. They were bound for Glasgow, where Mr. Bostock's brother ran a zoo. Indeed, in Scotland, the newspapers were already drumming up attendance, telling readers that Chiquita, "the sensation of Buffalo," was on her way.

It took less than two weeks for Tony to give in. Saying he was desperate to see his wife, he wrote to Bostock to apologize and said he was ready to join the show. "Apparently you have realized how foolish you have been," Bostock wrote back. He told Tony to report to his representative in Boston, where he would have tickets available for travel to Scotland.

On March 3, Tony said good-bye to his family and shook the hands of the reporter who had so faithfully reported his story. It would be a long good-bye, he said. "I go to join my Alice!"[13]

But there were no tickets to Scotland for Tony, and all he could do, once more, was wait. The couple did not reunite until mid-June, after Chiquita had performed throughout Europe and sailed back across the Atlantic. It was seven months after their marriage.

Bostock, meanwhile, bought $250,000 worth of life insurance for Chiquita. It was an expensive policy, but he drew on her salary for half of it. The showman then hired two attendants, Mr. and Mrs. Charles Badger, to oversee the couple as they toured with one of his shows.

Alice Cenda had endured harsh treatment from Bostock over the previous year, but nothing beat the soul-breaking rule of the Badgers. As the show moved through the South and the Midwest in the summer of 1902, the Badgers made sure the Animal King got his money's worth out of his Doll Lady. They withheld her eighty dollars a month. They refused to buy her street clothes, forcing her to wear the same calico dress for weeks at a time. They demanded she hold fifty to sixty receptions every day, from morning until midnight. They gave her almost no time to eat. Soon enough, she was worn out and losing weight.

On trains between towns, Tony tried to comfort her and brought her food. Another performer watched him give her "friendly attentions such as were customary from a husband to a wife." The Badgers didn't like it. Nothing if not conscientious, they eyed them closely. They locked the couple in hotel rooms. When Tony and Chiquita were allowed on the street, the Badgers stayed close. To make matters worse, Mr. Badger drank, and when he did, he swore at Chiquita and was "abusive." At times, Chiquita said, "it was almost unendurable."

But then came Elgin.

On Sunday, August 24, 1902, the Badgers and all the Bostock performers arrived in the Illinois town of Elgin, northwest of Chicago, for a one-week show. As they did at all the places they planned to stay, the Badgers surveyed their living quarters. Mrs. Badger made it clear to Mrs. Page, the proprietor of the boardinghouse, that she needed a room close to Tony and Chiquita. "These freaks are strange people," she said, "and we have to watch them so they will not run away."

The group settled in, and while Chiquita held receptions in a show tent, Mrs. Badger and Mrs. Page sat in the parlor, chatting. Taking the landlady into her confidence, Mrs. Badger explained how important it was to keep Chiquita afraid of her. In fact, she admitted, "she often had to whip [her]." She had also taken pains to keep the couple isolated. Sometimes Alice had been invited to people's houses during their tour, but "it was Bostock's orders that she be allowed to visit no one." She also confessed that they wanted to eliminate Tony. Thus far, they "had tried every way to get rid of him."

Maybe Mrs. Badger irritated the boardinghouse owner. Or perhaps Mrs. Page, seeing a small, beaten-down wife and her fretful husband, felt sorry for the couple. However it happened, the landlady told Tony and Alice that their overseers were concocting a plan to separate them, "even if they had to do away with [Tony]."[14]

August 26 had been a busy day in Chiquita's tent, so in the rooming house, the performer retired upstairs with her husband. Mrs. Page went to bed. The Badgers, their surveillance done for the day, headed to their rooms, too. Time passed. The town stilled. After midnight, Tony looked out the window, waiting for a sign. He had telegraphed Erie friends, praying that they would drop what they were doing and travel across three states to help him out.

He spotted them. Down on the street, the men waited with a carriage. Tony opened the upper-story window. He wrapped a long rope around Alice—how he got the rope, nobody knows—and carefully, soundlessly, lowered her to the ground. He followed, then hurried to the carriage. Neither he nor Chiquita took anything with them, not even his prized cornet. Chiquita was in her nightdress.

The friends whipped the horses, galloped out of Elgin, and pulled up at the Geneva railroad station, twelve miles south. Carrying Chiquita like a tired child, they took an early morning train to Chicago, arrived in the city at 9 a.m., and slipped out of the station. Worried that the Badgers or Bostock might have alerted the Chicago police, they didn't dare take a cab across town to the depot for the Lake Shore line. Instead, they hailed a drayman on the street and climbed into his wagon. They rode openly, "trying hard to look innocent and like farmers."

In Erie, the reporter at the *Erie Morning Dispatch* got a message that the couple he had dogged so faithfully was free. Hustling to the Woeckener house, he found Alice still in her nightgown—and ready to talk. Chiquita told him about the Badgers, and then one of Tony's friends took over the tale, describing their escape. "No one in all the crowds paid any attention to us," he explained. "They took [Chiquita] for a little baby I was carrying." As she listened to him tell the story, Chiquita, probably for the first time in months, laughed.

The newsman was beside himself. He imagined how Bostock must have looked when he found out that Tony and Chiquita had escaped— how he probably "tore his hair and filled the air with blue blazes." He couldn't believe that the man had been outsmarted. "Have they really escaped from the mighty Bostock at last?" he asked. "Is it possible that the man who in all his long show career has never before been 'done,' has at last been outwitted?"

Not quite.

Bostock's men assured the public that Chiquita would soon be found. She might miss some shows in Elgin, they said, but she would appear in South Bend, Indiana, her next scheduled show. "Oh, we'll find her all right," commented one. They told the press that she was a thief, too—that she had stolen all of Bostock's jewels.

The Animal King put detectives on the case. Years later, Tony's brother Eddie remembered that "they hid behind the trees out front of [the house] trying to catch them." But this time, Tony's whole family turned out to guard them. Even his grandfather, in his seventies, came from upstate New York to pitch in for three weeks.

Bostock threw more troops into the battle. Learning that Tony and Chiquita were making plans to stage a show of their own, he sued the performer to prevent her from breaking her "iron-clad" contract. He was likely furious to see the publicity they created, too. Their advertisement for the Edinboro State Fair in Pennsylvania described Chiquita as "the smallest woman on earth, stolen by Bostock and kept in captivity for the past year."

Through the fall of 1902, the court fight in Pittsburgh moved forward, with Bostock calling in witnesses like Jack Bonavita, the lion trainer, and Chiquita and Tony drawing on the testimony of Mrs. Page, the Elgin boardinghouse owner. Even the Buffalo justice of the peace who married the couple was asked to provide testimony.

Bostock lost. The case wasn't formally settled for three more years—the details have disappeared—but in fact Bostock gave up. Chiquita was finally free. In January 1903, when she performed again, she did so with her husband. She held a "dainty reception" on State Street in Erie, where children could meet her for a dime. The notice of her show described her as "a lily with a heart of fire, the fairest flower in all the world."

Tony probably wrote it himself.[15]

11.

The Timekeepers

I

THE ELEPHANT IN THE CASTLE

While Tony and Chiquita sweated out the decision of the courts and hid away from Bostock in Erie, one hundred miles to the west, in Cleveland, Ohio, the Animal King played another hand, in his other game.

After exhibiting Jumbo II to New England crowds through the winter of 1902, Bostock shipped the elephant off to Baltimore. There, Jumbo broke out of his pen and tried to find his way out of the grounds. He was recaptured. The showman then hauled him to Cleveland and tried to sell him. Having billed the pachyderm as the "most savage elephant in the world," he couldn't have been surprised when his offer to the Wade Park Zoo was turned down.

His sales efforts having failed, Bostock positioned his big elephant as a curiosity at his Cleveland show. Forty horses carried him in a wooden cage to Manhattan Beach, at the edge of Lake Erie, and there, throughout the summer of 1902, in the company of his usual mates and alongside Madame Morelli, Captain Bonavita, and the formerly mad Professor Weeden, he held court in a "castle." His biography was modified to match his new quarters. Not only had Jumbo caused the death of "hundreds" during his life, but in India he had been "used as a public executioner and has crushed the lives out of hundreds of criminals beneath his massive feet." Most recently, at the Pan-American Exposition, he had marked his arrival "by killing a man." He was "the worst elephant in the world," and all it cost to see him was twenty-five cents.

Through the summer on the waterfront, Jumbo stood in his castle, as Big Liz and Little Doc gave rides. It had rained heavily in June, and Ohioans, released from indoors, poured into the menagerie. Crowds particularly liked feeding time with the elephants. With the exception of one incident when Jumbo roughed up a new employee who had entered his stall and teased him, he was most "reasonable."

In September, after the children went back to school and the crowds thinned, the show on the beach closed. Stage sets were struck, cages were dismantled, and Bostock's animals were loaded onto trucks and wagons and taken to the train station to move to winter quarters. Big Liz, Little Doc, all the other trainers and animals—everybody left Manhattan Beach.

Everybody but Jumbo II.

A man in Baltimore, claiming Frank Bostock owed him a lot of money, launched a lawsuit. Jumbo was attached to the suit, and a helpful Cleveland sheriff ordered Jumbo held until Bostock paid up.

Bostock left town and disappeared and the sheriff rued his decision. The elephant, he discovered, cost money. His oatmeal, peanuts, and bales of hay were expensive. A hired elephant keeper cost more than thirty dollars a week. "Something must be done about this pretty soon," said the sheriff in early October.

In the middle of the month, the sheriff again complained that Jumbo's daily needs—two bales of hay, barrels of water—were too much for the county budget. Bostock had been tracked to New York City, but he had not replied to inquiries.

Finally, and probably under pressure, Bostock settled the suit. Jumbo II was his again. And within days, the showman made an announcement. A bullfight manager in Mexico City wanted to buy the elephant to fight a bull in his show ring. Even better, he would pay Bostock $5,000 for his big pachyderm. The Animal King, who had been plotting for months for ways to get rid of Jumbo, now made it clear that he had every reason in the world to keep him alive.

Which is why it was an unfortunate coincidence when suddenly, and for real this time, the big elephant died.

Some said that Jumbo II had been poisoned. Others said that the elephant, left by himself in the cold amusement park in Manhattan Beach, had died of loneliness. He had been sick for days, they claimed, and was "longing to join his companions of the show." What is certain is that someone who knew exactly when and where the elephant would die, and who knew exactly what sort of equipment to bring, quickly sawed off the elephant's tusks. The "thief" was never apprehended.[1]

Jumbo II never stood a chance against the Animal King. But at least his menagerie mate Big Liz carried with her some news that softened—if only slightly—the fact of his untimely end. After Cleveland,

Big Liz and a baby elephant take tea at Bostock's, at the Buffalo fair.

Liz was moved on to shows in New York City. And on June 24, 1903, between 4 and 4:30 p.m., at the Sea Beach Palace in Coney Island, she gave birth to twins, weighing approximately 150 pounds each. To those who did not know the recent history of the elephant mother, the *New York Times* offered some help. "The father of the babies is Jumbo II," the paper explained, "whom Liz met at the Pan-American Exposition twenty-two months ago."

Frank Bostock established Coney Island as his base for the next few years, opening a show in a brand new amphitheater and staging performances for fifteen thousand people daily. He continued to exhibit elephants, including one named Blondin, who in 1908 walked a tightrope and smoked a pipe. Many of his trainers, including Jack Bonavita, remained with him. Bonavita lost an arm to a lion in 1904, and then, following an attack by a polar bear in 1917, his life.

Bostock continued to tour the United States through the 1910s, and also took his acts overseas to Cuba, Europe, Australia, and South Africa. He expanded his business to theaters, ballrooms, and skating rinks. Then, suddenly, in London, in October 1912, Frank Bostock died. The cause of his death at forty-six, it seems, was influenza, and his funeral, by all accounts, was grand. Floral tributes arrived from business partners around the world—wreaths with a lion's head, a floral kangaroo, a life-sized lion. Carriages to the Abney Park Cemetery north of London brought friends and admirers. Later, his family erected a tombstone with a big marble lion, recumbent and sleeping, for his resting place.

In memoriam, the press talked about the practices of the "Great U.S.A. Bostock," as he was known. He was, the *New York Times* said, a man with a talent for staging remarkable publicity stunts. Others stated that his courage with wild animals was legendary. Still more talked about his theories on animal training. Addressing rising public

Bostock's longtime lion trainer, Captain Jack Bonavita, ca. 1903.

concern over animal welfare, they said that Bostock delivered correction as a last resort. And parroting decades of press releases, they claimed that his creed was "kindness, kindness, kindness."[2]

Through the twentieth century, elephants like Jumbo II and Big Liz continued to serve as public sensations. And, just like Bostock's elephants, they were viewed with mixed emotions. Shortly after Jumbo II was killed, a circus elephant named Topsy was electrocuted at Coney Island, in front of thousands of adults and children. (An Edison motion-picture company filmed the event.) The press, weighing in on the event with the same blend of prurience and remorse it had shown at Buffalo, described Topsy as both an incorrigible man-killer and a docile pet. She came to her killing site as "gently and obediently as a child."

Theodore Roosevelt provided the American media with more elephant stories (and more ambivalence) when he publicized his East African safari in 1908. While collecting animal specimens for American museums, he disparaged Native African hunters for their lack of sophistication, boasted about Western guns, and marveled at elephants. They were terrifying beasts, he claimed, and he filled his account with stories of maulings, gorings, and tramplings. When he killed one, he stood on top of the world. "I felt proud indeed as I stood by the immense bulk of the slain monster," he wrote after a shoot, "and put my hand on the ivory." After killing one pachyderm, he roasted pieces of its heart and "found it delicious."

Roosevelt saw in the elephant something more than terrifying majesty, though. The elephant was intelligent, savvy to the ways of the hunter, and skilled at avoiding them—it was what made the animal such attractive prey. They were not only wise; they were sociable. It was fortunate that there were so many of them, said Roosevelt, and that there was no danger of extinction.

In the United States throughout the twentieth century, the reputation of elephants continued to soften. In parade after parade, circus after circus, they plodded their way into the public's affection. They sat on balls, balanced on tubs, were laughed at and admired. Then in 2015, the Ringling Bros. and Barnum & Bailey Circus announced that it would retire its herd. The decline of elephant populations worldwide and the recognition that these giant animals might be gone forever played a part in the decision. But more effective than that were the protests of animal advocates, who publicized the way they were imprisoned, disciplined, poked, and pricked. They also emphasized that these mammals had a humanlike way of solving problems and expressed fondness, grief, and empathy.[3]

What took place in 1901 in Buffalo—where Jumbo II pushed against his confinement and suffered the consequences, and where the brand new SPCA fought to prevent a spectacle of animal killing—pointed toward this shift in thinking. Jumbo's "moment" did not mark any watershed in public opinion, but it did reveal a dent in the belief that civilization meant the subjugation of wild animals, and it served as a sign of changes to come.

II

EVER AFTER

Released from Frank Bostock's dominion, Tony and Alice Woeckener spent the winter of 1903 dreaming up new shows. Tony corresponded with "several little people" and considered putting together a traveling troupe, accompanied by the Woeckener family musicians. Best of all, a writer was working on a pantomime that would feature special scenery and star Chiquita. For once, the performer would headline a show that was not her personal drama.

Chiquita stayed in the Woeckener home for much of that winter, resting and eating. She gained weight. And she became pregnant. The event, not surprisingly, made national news. Nine months later, however, in mid-October 1903, Chiquita signed her last will and testament and was prepared for a caesarean section at St. Vincent's Hospital in Erie. The surgery was described as "a last resort to save her life." The operation did end up saving her life, but her son was stillborn.

Over the next decade, Chiquita, with Tony as her manager, continued to tour. She performed in the West, in Mexico, and throughout Europe. One visitor who saw her show in New York City in 1909 said that just as endearing as Chiquita was her proud and attentive husband. He carried her to the platform, and then watched her act. "His devotion to his, the smallest wife in the world, is seemingly unending."

It was in the spring of 1928 that a Buffalo newspaper announced that Chiquita, who had returned to Mexico with Tony sometime earlier, had died in Guadalajara. She would have been around fifty years old. The paper knew that many of its readers would have no idea who Chiquita was, so it reminded them. She was the young woman, the paper explained, "whom gaping hordes milled and jostled to see when she appeared at the Pan-American Exposition in Buffalo in 1901." The announcement said that Chiquita's end "brought her merciful relief." The choice of words suggests that she might have had a hard time toward the end of her life, but it is comforting to imagine it was just a

Tony Woeckener and Alice Woeckener (Chiquita), ca. 1904.

turn of phrase. What is evident is that her last decades with Tony were good. The newspaper recounted the story of their romance in Buffalo and claimed that they lived happily ever after.[4]

World's fair midways continued to be popular venues for displaying short-statured people like Chiquita. In the 1930s, it wasn't just one Little Person who was featured; expositions hosted "midget villages" and "midget circuses." By the late 1900s, though, individuals with extraordinary bodies were increasingly viewed as medical anomalies that needed to be fixed. The hospital room, in other words, picked up where the circus tent left off. At the same time, and partly as a result of being set apart for so long, Little People joined together to demand rights, respect, and, above all, a reconsideration of what a normal body really meant. Alice Woeckener, with Tony's help, had staked a claim to freedom on her own. A century later, she would have had the support of a wide and loyal community.[5]

III
THE OLD ADVENTURER

When Rainbow City locked its gates for good on November 2, 1901, the head counters went to work and calculated that the Buffalo Exposition had seen more than eight million visitors. It wasn't the twenty million the big talkers had predicted, or the ten million that would have set the accounts straight. It was, all things considered, a respectable number.

Almost immediately, frets about attendance were repeated farther south, when another show opened. On December 1, 1901, Charleston, South Carolina, welcomed visitors to the Interstate and West Indian Exposition for a six-month run. The fair's designers eschewed color this time and went back to tried-and-true white. The fair, located on the grounds of an old Ashley River plantation, became known as the "Ivory City." It recorded 675,000 guests.

Charleston fair directors brought in some Buffalo shows. The Cuba exhibit was reassembled, as were The Streets of Cairo, Darkness and Dawn, and the Esquimaux Village. And, in the opening-day parade, who should bring up the rear but elephants, lions, and zebras from one of Frank Bostock's menageries.[6]

Another familiar face appeared, too: the barrel-riding star of Niagara Falls. The days since her bandstand appearance at the Pan-American Exposition had not been easy for Annie Taylor. While Tussie Russell booked engagements for her in early November, she had fought exhaustion. She persuaded Russell to take her back to Bay City, but even there, Annie said, Russell rented a storefront and "dragged me there more dead than alive." She also complained that he didn't give her all the money she made.

There had been other problems, too. Her story had made its way around the country, and, in Michigan, a blacksmith named Montgomery Edson had scanned the news about her exploit with more than usual interest—and perplexity. He became so deeply curious about the matter, in fact, that he decided to make himself known to the press.

He was, he said, Annie's brother. There was only one set of Edsons who lived near Auburn, New York, and the details of her life fit with what he knew of his long-lost sister.

There was only one thing that didn't match up. If Annie Edson Taylor was his sister, then she wasn't forty-three years old. In fact, the person who had just dazzled the world with her nerve and daring was almost an old lady. And all those witnesses who said that she looked her age, and wondered whether she were past her dancing prime? They were right.

While papers in Michigan buzzed about Taylor's age, Tussie Russell, who was taking offers for shows, kept silent on the subject. Annie, too, fired back. When a Bay City reporter asked her whether she was Montgomery Edson's long-lost sister, she had a sharp answer. "He's a fool," she said. "I'm nobody's lost sister and I never was lost." But she did not deny the relationship.[7]

In mid-November, when Taylor began snapping at reporters and shooing away spectators, Russell drove her to a sanatorium for a rest. Several days later, the two were back on the road, holding court in Flint, Saginaw, Detroit, Cleveland, Cincinnati, and Columbus. In most of these cities, Taylor sat in a department-store window, her cat (or, more likely, *a* cat) and her barrel by her side. Occasionally she gave a short talk.

They made almost no money. In Cincinnati, Annie explained, one of her fellow Niagara Falls stunters, Martha Wagenfuhrer, had arrived first, captured Annie's audience, and "imposed on the public." Annie tapped into an old Texas bank account in order to get to Charleston, and, after appearing at the fair for two weeks, she and Russell couldn't even buy train tickets out of town. They turned to public-assistance offices for help.

Russell blamed their failures on Annie's "peculiarities" and on her looks. "I've trotted [sic] her around the country," the manager said, "but she got to be a frost. . . . People took more interest in the barrel and the cat than they did in the woman . . . they didn't believe she was the woman." He explained that audiences had imagined her as a youthful "damsel," and that when they came face-to-face with the adventurer and found a woman "somewhat down toward the sunset side of life," they took no further notice of her. They could have profited, he said, if she had been "a beautiful girl."

With these ideas in mind, Russell made a break for it. Commandeering the barrel, he sold it to a Chicago theater company that wanted to produce a play based on Taylor's ride. The manager of *Over the Falls*, as the show would be titled, suggested he might even have a part for Russell, either as hero or villain. He would not only feature the original barrel but also reimagine the heroine as lovely and young.

While *Over the Falls* went into rehearsal—according to theater circles, it seemed destined to be a hit—the barrel's original rider struggled on. An appeal to the Ohio Relief Department in Cleveland gave Annie funds to go east and, in the winter of 1902, she returned to Niagara Falls. Desperate for money, she marketed postcards and produced a hastily written memoir. In late March, she wrote to a Niagara Falls theater manager for help. Tussie Russell, she said, had a "villinous [sic] temper," and had proven himself rotten. He had taken her livelihood and made it his own. "Despair as bitter as death," she said, "has settled down on me."[8]

As money from souvenir sales accumulated, Taylor's attitude shifted to determination. She hired a Chicago lawyer named Moody to find her barrel, and Moody, in turn, hired a detective. On August 14, he found it. The cask was brazenly being shown in a Chicago department-store window advertising *Over the Falls*.

Taylor made plans to go to Chicago immediately. Worried that the theater managers would do everything possible to thwart her, though, she decided to travel under an alias. Mrs. Isaac Davy, the name she chose, belonged to an acquaintance. She took with her a decorated alligator bag, so Moody would know her right away.

In Chicago, she and the lawyer, with the help of city police, moved fast. As a driver backed a truck up to the department store, an officer served papers on the manager. Moody's assistants maneuvered the barrel into the truck and drove away. Twice that night, they switched the barrel's location, and, by morning, they had the cask on an eastbound train, in the company of Annie Taylor. She was delighted.

Through early fall, Taylor toured northeastern state fairs with her prized possession. With her new manager, William Banks, she sold her memoir and photographs and spoke to audiences about her ride.

At the end of September, she left for a new engagement in Trenton, New Jersey. As usual, her barrel traveled separately.

There was something about Annie Taylor that tempted men into usurping her story. Instead of reconnecting with Taylor, Banks took the barrel, dressed up a woman named Maggie Kaplin in clothes like Taylor's, gave her Taylor's books to sell, and launched a tour. Annie went to the Trenton justice of the peace to report her stolen barrel and met with other lawyers. Then, depleted of money and strength, she gave up. She could read about her barrel making the rounds of theaters in "startling," "marvelous," and "dramatic" renditions of her descent, but she never saw the real thing again.

The remaining years of her life saw the aging adventurer trying

Annie Taylor sells souvenirs at a stand in Niagara Falls.

to market her feat with a replica barrel, and hawking souvenirs on the streets of Niagara Falls. She watched more theaters and filmmakers tell her tale and even toyed—briefly—with the idea of going over the falls again in 1906. Five years later, a forty-seven-year-old Englishman named Bobby Leach stole some of her thunder when he tumbled over Niagara Falls in a steel barrel. He was badly injured but survived. Taylor said she did not want to discuss the matter.

At the age of seventy-six, Taylor offered to portray herself in a movie about her pioneering plunge. She got into a barrel upstream of the falls and reentered it downstream. The production was never released. She dreamed up other schemes to make money, too, but now, nearly blind, she could not make them work. In the winter of 1921, she was admitted to an almshouse north of Buffalo. The registrar noted her "cause of dependence . . . sore eyes, no home, no money." Other officials also noted her age. Ever the optimist, she had told them she was fifty-seven.

The poorhouse did not diminish Taylor's willingness to give interviews about her adventures. Four trips across the Atlantic, she said, fourteen times across the Florida Straits, "several" times across the continent, and once deep into Mexico. And one other, of course. She also continued to reimagine her station in life. She told a newsman soon after she had been admitted that she was writing a historical novel she thought would be just right for *Harper's*. Then she could leave the almshouse. "It is quite a change for me to come here when I have been used to being entertained in senators' homes in Washington and travelling extensively," she said.

She died there that spring, on April 29. She was eighty-three years old.[9]

Annie Taylor never succeeded in entering the high society she admired, either in 1901 or beyond. She was too poor, too restless. She also failed to secure a livelihood from her descent of Niagara Falls. She was, everyone said, too old. There were other problems. When men "conquered" Niagara by barrel, they were regarded as brave, if somewhat foolhardy. When they harnessed the cataract through engineering, they were seen as performing a brilliant, powerful, and profitable accomplishment. Taylor's feat seemed to provoke as much irritation as wonder. Not only had she used an old-fashioned barrel to achieve her goal, but her act seemed to make the falls mundane. The "grandeur of the cataract," explained a Niagara Falls park superintendent in 1902, "has been largely augmented by the fact that its dangers have been forever invincible by human beings." When Taylor came along, he said, people "were resentful that anyone has triumphed over the mighty Niagara. They do not care to hear the story of a matronly appearing woman who has tripped up Hercules."[10]

The grand escapade that culminated in Annie Taylor's appearance at the Pan-American signaled change ahead for women, however. Women could vote by the time Taylor died, and women could travel alone with less suspicion. Eventually, women could enter the public sphere as politicians, professionals, and, thanks to the likes of Amelia Earhart, as adventurers. But, particularly if they were women of color, they would still contend disproportionately with poverty. And, for generations to come, older women would still be relegated to society's backseat. Like Annie Taylor, many would try to defy their age, or lie about it, in order to capture the value accorded to youth.

IV
NIAGARA

Like every exposition, the Charleston fair that proved so disappointing for Annie Taylor was a magnet for controversy. Women in Buffalo had thought that a separate pavilion filled with women's work was a backward step. Charleston women disagreed, and they proudly erected a building to show off women's handicrafts and literature, and offered a place where mothers could deposit their babies. The Charleston exposition also featured a Negro Department headed by Booker T. Washington and a Negro Building with displays, including the Du Bois exhibit, which demonstrated African American progress since Emancipation. The building was set off in the "Natural" section of the fair, as opposed to the Art section with most of the other main buildings. A number of black Charlestonians objected to the segregation of both the Negro Department and the building, and also took issue with the art. A group statue in front of the building featured a "Negress" balancing a basket of cotton on her head, a man using an anvil, and another holding a banjo. Its designer claimed it celebrated African Americans for their "great ability as tillers of the soil and mechanics." People of color in Charleston protested. They found the figures degrading and forced the statuary to another site on the grounds.

The Old Plantation concession from Buffalo's Midway did not go to Charleston. What played well among white fairgoers in New York State would likely have fired a storm of anger in a city that was fifty percent African American and who knew, all too well, the truths of plantation life. Instead, the show moved on to Coney Island.

His stint as Laughing Ben thus over for the time being, Ben Ellington went back to Dublin, Georgia, and made a triumphant return. Having earned three dollars every week in salary, and as much as five dollars a day in tips, he brought back four hundred dollars. He hoped to buy a small house.

Ben became a town treasure. He was called on to greet distinguished visitors to Dublin and continued to use his talents to make money and to navigate his way through hard times. He avoided prosecution for occasional theft and laughed himself out of court fines. When he couldn't produce identification, his laugh became his signature.

Ellington adopted a persona that worked for him. In a world of segregation and lynching, where black men met oppression at every turn, his performance as an inoffensive, happy man must have seemed a strategy for survival. As one newsman who met him commented, the old performer, despite his belly laughing, was "nobody's fool."

After completing several more national tours, Ben Ellington performed for the last time during the Louisiana Purchase Exposition in St. Louis in 1904. He died back home in Laurens County, amid the red dirt hills he loved, in 1905.[11]

While Ben survived the Jim Crow era using his laugh, Jim Parker, who had tackled Leon Czolgosz in the Temple of Music, took his Pan-American fame on a different route. After hearing so often that he had been a "credit to his race" by aiding the stricken president, and after rejecting offers to go on stage in shows, he took to the lecture circuit.

In late December 1902, he reported receiving a new tribute. McKinley's friend, Senator Mark Hanna of Ohio, wanted to do something to make up for the way that Secret Service men and others at Czolgosz's trial had dismissed Parker's brave act. Hanna offered him a position as messenger in the United States Senate.

If Parker took the position, or held it long, we do not know, but for the next three or four years he continued to speak, often under the auspices of the AME Church. A handbill printed in Worcester, Massachusetts, spoke of him as "The Greatest Negro of the 20th Century." Despite such labels, or perhaps because of them, Parker struggled for credibility. His efforts to be believed for what he did in Buffalo became a mission, and the mission was clouded by illness. In Atlantic City, in the spring of 1907, he was "arrested in a demented condition" and charged with vagrancy. The press maintained that he had gone from street to street, telling his story to passersby and crowds that gathered. White commentators suggested he had been unable to handle the accolades that had been bestowed on him and had succumbed to "fast living."

Hospitalized, Parker persisted in appealing for help from the people who knew him from Buffalo. Sometime in 1907, he penned a letter to Ida McKinley:

> Kind Madam. I write you to ask a favor of you. I have been sick in the hospital for some time and am sick yet. I want you to help me. I done all I could for your husband in trying to save his life and if I had of been successful I know he would of [found a place for] me for my efforts a word from you to Mr. Cortelyou is all I need.

Parker said he would be happy for anything, from a place to live to a job driving a wagon. Please, he asked, "do this for me in remembrance of your dear husband."

Ida McKinley asked George Cortelyou—who by then was serving in the cabinet under Roosevelt—to help Parker out with a job in Washington and a home in the district. Whether or not Cortelyou did something for Parker is unclear. What is certain is that Parker's life did not turn around. In the winter of 1908, the police again picked him up, this time on the streets of West Philadelphia. He was admitted to the "insane" department at the Philadelphia Hospital, and in late March, at age fifty-one, he died. Having no connections in the city, his body was given to Jefferson Medical College, and, two weeks later, was set out on a dissecting table for the benefit of medical students.[12]

While James Parker died institutionalized and alone, other people of color who had been active at the Buffalo Exposition saw more enduring success. The local community that had brought the Du Bois Negro Exhibit to the fair did not let up. Mary Talbert, in fact, figured in another important story. In 1905, W. E. B. Du Bois wrote to Mary's husband, William Talbert, about finding a "quiet place" for a top-secret meeting near Buffalo. The meeting would discuss strategies to counter the accommodationist policies of Booker T. Washington and to demand equal rights for African Americans. In early July, a group of men, including Du Bois, met first at the Talbert home and then at a hotel in nearby Fort Erie, Canada.

Barred from the meeting, Booker T. Washington asked his wife to write to Mary Talbert in Buffalo, to keep him "closely informed." He also hired local men to spy on the gathering, and he tried to block newspaper coverage of the event.

He was right to be concerned about the clout of the men getting together. Du Bois and the others, who spent two days in discussion, were launching what they called the Niagara Movement. At some point in their deliberations, in fact, they made a short excursion to Niagara Falls. What Du Bois made of the falls at the time, he didn't say, but later he described the cataract as one of the "wonderfulest" things he had ever seen, and he attached its name to his new organization. Four years later, the alliance broadened, its leaders joined with other activists, and it became better known as the National Association for the Advancement of Colored People. The twentieth-century fight for civil rights was up and running.[13]

V

THE CENTER OF THE UNIVERSE

On July 1, 1902, after a week of gloomy weather, the sun emerged and, for the final time, struck the gilded iron statue on top of the Electric Tower with a bouncing light. A workman from the Chicago House Wrecking Company climbed the tower, fastened a rope around the neck of the Goddess of Light, and ordered men to pull. The statue somersaulted twice, plummeted into two feet of mud at the Court of Fountains, and broke into pieces. Witnesses at the scene were dismayed but not entirely disappointed. The statue had been sold to the owner of a popcorn pavilion in Cleveland, and, had the buyer appeared in time, would have had an ignominious end. "Cleveland, of all places—a popcorn joint, of all thrones," lamented one journalist. Better, he suggested, to have her deep in the Buffalo dirt.

Rainbow City shut down in the first week of November, 1901, but it disappeared slowly. Unlike Chicago, which saw much of the White City go up in a conflagration four months after its gates closed, Buffalo lost its fair in increments. For almost eighteen months, residents watched buildings lose their plaster façades or go down with the wrecker's ball.

The Midway was the first big section to go. By the spring of 1902, it was little more than a wasteland. Where once spielers, brass bands, and Bostock's animals had filled the air with noise, now only sparrows cheeped, pecking away at grit and dust.

The buildings on the formal courts were brought down more gradually and stood as skeletons for months. Next to them, and in nearby Delaware Park, the ground was piled high with twisted metal and wasted wood. Those who visited the site said it looked like a hurricane had hit the area, while others wouldn't visit the grounds at all, saying it was just too sad.

By the summer of 1902, the grounds were empty. A reporter, now relegated to covering humdrum life, looked back at the previous year. "You miss the Exposition if you're a Buffalonian," he wrote. Some parts of the fair seemed so completely erased, he added, "it is mostly a dream."

Buffalo residents had tried to save some of the dream. Even before the Exposition's end, city dwellers had plaintively wondered how they might hold on to it. "Must all the beauty and magnificence of the exposition be thrown away?" asked the *Express*. "Ten thousand times has the question been repeated in this city." The loss of the beautiful

The Triumphal Bridge in ruins, June 1903.

buildings had weighed on visitors during the fair's last days, the paper said, and robbed them of pleasure. The Exposition seemed like a fairy-tale beauty, "condemned to a horrible death by the giant ogre."

Ideas for saving some of the site poured forth. A local man suggested the grounds become a national park, because the president had been shot at the location. New Yorker John Carrère, the fair's architect, threw out a different plan. The beautiful Pan-American fair had cost a lot of money. Why not keep the canals, the fountains, and even the Triumphal Bridge and the Electric Tower? Why not plant new trees and make use of the vistas? Think, he said, of "Versailles and Fontainebleau." It would be an investment such as they make in Europe. If Buffalonians would only show the same "public spirit" they had earlier.

Carrère wanted Versailles. Others said they would be happy with a toboggan run. They pictured a mammoth slide running from the Electric Tower to the Triumphal Bridge, and then down the incline of Park Lake. The canals and fountain basins would become skating ponds, lighted at night. A concessionaire had an even more extravagant thought: Why not spray water onto the Exposition buildings and turn them into fanciful ice palaces? A local businessman thought the "scheme would, I venture to say, draw people from nearby cities."[14]

But officials had had enough of such schemes. Who could dream of a new spectacle when they had not yet paid for this one—when, as one commentator put it, "the surplus was on the wrong side of the balance sheet?" The fair had cost close to $7 million and had brought in $6 million. Many contractors and laborers who had built the Exposition buildings had not been paid at all. Mortgage holders had been reimbursed, but not completely. Stockholders would see nothing.

The Exposition Company and all the people who had poured hearts, souls, and money into the fair hoped that Washington would help out with a million dollars. The federal government had not helped fund the fair, after all, as it had done for other big expositions; it had simply paid for its own exhibit buildings. And hadn't the fair failed because it had hosted the president, with disastrous results? One newsman summed up the case: "The fact that the President's death occurred in Buffalo at the time when the full tide was welling and that it stopped the growing throng and broke the thread of enthusiasm, gives the Exposition Company a certain amount of claim on the country at large." Exposition directors brought out attendance numbers and projections that backed their claims—McKinley's death, they said, had cost the corporation $1.5 million.

The fair's movers and shakers knew, of course, that the Exposition's losses were grounded in more than the death of an American president. They had had worries about low attendance back in the early summer, when immense crowds weren't materializing. They had blamed the numbers on the weather: Spring snows and summer chills had delayed construction and dampened the enthusiasm of early visitors. They also thought that railroads hadn't lowered rates soon enough. Concessionaires had other opinions. Closing the Midway on Sundays—turning the Exposition into "a funeral," as some of them put it—turned off patrons. Others thought the advertising was off. Pan American promoters, they complained, did silly things like print ads "on miniature frying pans and beer mugs or in pretty booklets and folders." If they had put an ad in a thousand daily papers, and run it for three months, it would have been far more effective.

Maybe, though, it boiled down to expectations that were too high. Looking back, William Buchanan described the attempt of a city of 350,000 to hold an international exposition without the financial backing of the nation or the state as a "hazardous undertaking." And Marian De Forest, secretary of the Board of Women Managers, said that the city had held out too much hope for patrons across the Americas. "Those most familiar with exposition history," she wrote, "have no hesitation in saying that it was too large for a city like Buffalo."[15]

John Milburn, who would travel to Washington and ask for help, would never say such things. He reminded people that the fair's officials—unlike the press—had really only hoped for ten million visitors, and the assassination had unraveled the rate of attendance. He also couldn't let pity ruin the pride the city felt in hosting the Pan-American Exposition. He walked a fine line—he had to let the country know the Exposition Company needed help, but he needed to remind Buffalonians that they had done a magnificent thing. To local residents who had lost money, Milburn pointed out that Buffalo had just attracted a slew of new investors and that millions of people had emptied their pockets in the city. If anyone dared complain that the fair had not matched the success of Chicago, he asked them—how convincingly is unknown—to think again. "The pro rata expenditure at the Chicago World's Fair was something like $14 for each visitor and as we have had a considerably better class of people . . . it is reasonable to suppose that the amount each visitor left here was at least $20."

Buffalonians of less wealth had invested in the fair, too, and to them Milburn offered comforting—if financially unhelpful—words. They had had "their ideas broadened." Furthermore, their hometown was now *known*. The Exposition, said its president, "has put Buffalo on the map of the great cities. It is no longer thought of as the town of which Grover Cleveland was once Mayor, but as the beautiful city on the lake where the Pan-American Exposition was held." Throughout the previous six months, important visitors had admired the city's fine homes and office buildings, its well-designed parks, and its asphalted avenues.

John Milburn also reminded people that the fair's theme had been successfully executed. "We have not lost sight of the fact," he said, "that the Exposition was held primarily to stimulate commercial relations between countries of North and South America. . . ." This goal, he announced, had been met, and, offering some selective evidence, remarked that "the English and Spanish languages mingled as they have never been mingled before. . . ." Other observers echoed the president. The Pan-American fair, said one, has been "a great loving cup from which all the nations of the Western Hemisphere have sipped." Chile had traveled seven thousand miles to take part in the big show, and Peru, Ecuador, Honduras, and Cuba had joined in "as if the Spanish language never existed."

Given the many translators necessary for speeches throughout the season, the Spanish language seemed alive and well. But the "Pan

American movement," as Milburn and others called it, in which Latin American republics opened themselves to investment and trade, and the United States took the lead in development, was considered well begun. It wouldn't be an American empire in the sense that the Philippines had now become part of an empire, but it wouldn't be far off, either.

Sounding like their old optimistic selves, local newspapers took up the beats of praise. Not only the country, but whole continents now held the city in esteem. The Pan-American had placed Buffalo "among the great commercial centers of two continents." Even more than that, the Exposition "has caused the eyes of the world to be fixed upon the Queen City." The magnificent Pan-American Exposition had even made Buffalo "on many occasions shine as the center of the universe."

It was early June all over again.

Pleas for help from Washington found a sympathetic ear. On the same July day that the Goddess of Light lost her noble perch, the United States Congress passed a relief bill giving $500,000 to the Pan American Exposition Company. John Milburn's months-long appeal had resulted in a whittling-down of the requested amount, but the money would help. Now, finally, the contractors who had cleared the grounds, built the Exposition halls, and prepared the grounds would be paid. Buffalo had other New Yorkers to thank for the rescue money. In the final vote in Congress, New York legislators had stood up for their western counterparts and "fought for Buffalo like Trojans."[16]

VI
MCKINLEY'S GHOST

The make-believe world of Rainbow City, with its minarets and arcades, its starry lights and luminous colors, did not last. It was not remade into an amusement park or a historic site, or turned into majestic gardens. Some bits of it, though, had a second shot at life. The floor of the Temple of Music, where the president had been attacked, was sawed up and stored in an office, and officials announced it would go to a museum. Bostock's big animal stage became a dance pavilion. The Exposition Hospital, although picked apart by souvenir hunters, was turned into a storage area for the wrecking company. In a reception area near where surgeons had treated the injured president, between seven and ten workhorses were housed temporarily, along with their troughs and feed boxes.

A more elegant future was in store for the New York State Building. Marbled and columned, more reminiscent of Chicago's neoclassical White City than the Spanish-style structures of the Pan-American, it became the permanent residence of the Buffalo Historical Society. Within its archives stand vast records of the Exposition, including, on a sturdy shelf, Mabel Barnes's record of her thirty-three trips to the fair. Her neatly penned and decorated homage to Rainbow City, in three volumes, took her from 1905 to 1914 to complete. Mabel continued to serve Buffalo schools, first in the classroom and then for thirty years as a librarian. She died at the age of sixty-nine, in 1946.[17]

The men who had led the Exposition and fought for it in the throes of crisis went separate ways. William Buchanan, director-general, continued to represent the United States in Latin American affairs. Fondly known as "the diplomat of the Americas," he brokered discussions and agreements in Argentina, Panama, and Venezuela. He continued his work in international business enterprises as well, marketing free enterprise to his friends in the Southern Hemisphere. His family was still living in Buffalo, and he was actively involved in diplomatic work in London in 1909, when, at age fifty-seven, he abruptly died.

His body was shipped back to Buffalo for burial, and Pan-American officials served as his honorary pallbearers.

Buchanan's partner in Rainbow City, John Milburn, tried to return to some level of normalcy in the fall of 1901, but it wasn't easy. Even after attendants removed oxygen tanks and electric fans from McKinley's sickroom on the second floor of his home, Milburn contended with reminders of the president's death. He now lived in the most famous house in Buffalo. Souvenir hunters took stones from his driveway and leaves from his trees. "Kodak fiends" surveyed the house from every angle. One man brought a chisel to the site to take away a few bricks. Touring coaches also stopped outside and guides used megaphones to identify the windows of the room where McKinley breathed his last. Some sightseers even asked to be admitted to the house.

In 1904, Milburn left Buffalo for New York City, where he carried on as a high-profile lawyer. He served as counsel to the New York Stock Exchange, and as a director of the American Express Company and the New York Life Insurance Company. He moved to an estate on Long Island, where his sons continued to play polo. Like his Pan-American cohort, William Buchanan, John Milburn died in London. It was 1930, and he was seventy-nine. His house on Delaware Avenue was torn down in the late 1950s.

Conrad Diehl had had enough of mayoring by the end of 1901 and left office just after the Exposition closed. Some said he was devastated by the death of McKinley—a painful blow to his treasured project. Diehl's time as mayor was marked not only by planning and hosting the Exposition but also by bringing electricity from Niagara Falls to operate streetcars and city lights. He also had to cope with a treasurer who ran away with more than $40,000 in city funds. After years of dealing with rich Buffalo businessmen, Diehl moved back into more comfortable work as a doctor and returned to his large practice. A robust man who had almost never been ill, he proved he was mortal on a stormy night in February 1918, when he slipped on ice during his evening walk and died a week later. "The Father of the Pan American," who was also known as "The Beloved Physician," was seventy-five years old.[18]

The man who was made President of the United States in Buffalo, Theodore Roosevelt, established one of the most energetic presidencies in American history. Pursuing some of the same concerns he had announced when he first set foot in the Temple of Music in Rainbow City, he attacked corporate wealth. He continued to promote the doctrine of manliness, as well as the well-being of Anglo-Saxons. He embraced Latin America, as long as Latin America did not engage in "chronic wrongdoing," and when it did, he embraced sending in the United States Navy to sort things out. (This policy, the Roosevelt Corollary, justified sending US troops to Latin America more than thirty times between 1898 and 1930.)

Roosevelt was also known for land conservation. Echoing the themes of the Pan-American Exposition, he asserted in 1908 that "America's position in the world has been attained by the extent and thoroughness of the control we have achieved over nature." But to maintain this position, and to encourage the out-of-doors hardihood he championed, he believed that wilderness—forests, mountains, and waterways—needed to be protected. As president, Roosevelt created game reserves, bird refuges, and national parks, altogether setting aside 230 million acres of land.

Roosevelt likely took many lessons from his days at the Buffalo Exposition, but embracing personal security was not one of them. He made a point of striding about Buffalo in the wake of the shooting, waving away worries about his safety. Six years later, on New Year's Day in 1907, he set a record by shaking hands with 8,105 people at the White House. Five years after that, his confidence, or faith in his constituents, or both, almost proved fatal. On October 14, 1912, while campaigning (unsuccessfully) in Wisconsin for a third term in office, he stood up in a carriage to greet a well-wisher. Another man took advantage of his exposure and shot him point-blank in the chest.

Roosevelt's folded campaign speech—it was a long one—helped shield him from a deep bullet wound, and he went on with his lecture. Even as his chest bled and his voice weakened, he persisted in talking. Only later did he agree to an X-ray, which showed a bullet in one of his ribs. It was never removed. John Schrank, the man who shot Roosevelt, claimed he had been inspired by a dream about William McKinley, who had appeared to him as a ghost and had spoken to him from his coffin.

The speech the wounded Roosevelt gave that night in Wisconsin had more than one echo of the shocking event in Buffalo. While continuing to remind his audience that he had been shot, and that he was all right, and that people could not use his injury as a chance to "escape" his speech, Roosevelt talked about what might have stirred up the would-be assassin. He blamed the "daily newspapers," with their "mendacity and slander which . . . incite weak and violent natures to crimes of violence." But he went further than that. "The incident that has just occurred," he said, signaled an ominous trend in America, where, if nothing was done, "we shall see the creed of the 'Havenots' arraigned against the creed of the 'Haves.'" If that day arrives, he said, shootings such as the one he just suffered would happen again and again. "When you permit the conditions to grow such that the poor man as such will be swayed by his sense of injury against the men who try to hold what they improperly have won," he concluded, "it will be an ill day for our country."

After William McKinley had been killed, no politician would have dared to make such an assertion. It would have sounded disrespectful, unpatriotic, and radical. It would have sounded too much like Emma Goldman, who, in response to McKinley's assassination, had suggested that Leon Czolgosz's violence was a result of "conditions," and explained that "there is ignorance, cruelty, starvation, poverty, suffering, and some victim grows tired of waiting."

Theodore Roosevelt did not readily champion the causes of working people. He worked from the top, not the bottom, targeting concentrated wealth held in trusts, banks, railroads, and corporations. Through his actions, though, and in rare speeches like this one, he did more than most presidents to address the inequalities of wealth at the turn of the century.[19]

Roosevelt never forgot to be kind to Ida McKinley. He sent wreaths to be put on her husband's tomb on Memorial Day and apologized if he went through Canton and did not stop to see her. For her part, Mrs. McKinley submerged herself in grief. She visited her husband's cemetery vault daily, cared for the fresh flowers that arrived, and wept. Edith Roosevelt, the new first lady, occasionally sent flowers to the site, and Ida McKinley sent them back after they had withered. In her house, she knitted slippers for charity and ate meals in silence. A friend visiting in 1902 said that the house, with its stale air and dead quiet, seemed like a "cemetery," and that its resident looked forward to her own death. "I only wait and want to go," Mrs. McKinley had lamented.

Contrary to general expectations, the death of her husband, and the grief that attended it, did not kill Ida McKinley. After several years of unhappy seclusion, she began to emerge. Her health seemed to improve. Her visits to the vault lessened, and she uncovered long-lost passions, such as her support for women's suffrage. Susan B. Anthony had earlier sent her a four-volume history of the movement, and nurses read the books aloud to her at night. She began to take an interest in theater and relished the company of her grandnieces—one of whom reminded her of her long-lost daughter Katie. In 1907, Mrs. McKinley suffered a bout of the flu, and, shortly afterward, a stroke. She died on May 26, 1907. President Roosevelt headed the government delegation at her funeral.[20]

VII
THE PROUD QUEEN

The 1900s looked like they would be good to Buffalo. The money that helped build Rainbow City paid for more industry, new blast furnaces and grain elevators and more railroads. The century that began with Buffalo celebrating the subjugation of a waterfall became the century that saw the city succeed in smelting iron ore, as it became one of the world's biggest centers for steelmaking. The Pan-American Exposition had done its part in honoring the transition when some of its lumber went into the construction of laborers' homes near the Lackawanna Steel Works. The Exposition also generated dozens of new enterprises and manufacturing plants, and thousands of new jobs. The Larkin Soap Company echoed other businesses when its executives said that exhibiting at the Pan-American had been "the best investment we ever made."

The Queen City boomed through World War I. During the next war and into the 1950s, there seemed to be no stopping it, as Buffalo's industrialists contracted to build automobiles, airplanes, ships, engines, and tanks. Some city boosters held onto the idea that Buffalo might match Chicago, even as Chicago spread wider and taller and became more populous. In the late 1920s, local advertisers maintained that Buffalo only needed better branding and more hustle. "Any other place in America with half the natural endowment of the Niagara Area," commented a local writer, "would be chasing New York for first honors and laughing at the other cities. Look at Cleveland—just a spot on Lake Erie, and not a very good spot, either. Look at Chicago—entirely surrounded by prairie and bossed by bandits—or St. Louis, with nothing from Nature except the stickiest, muggiest climate this side of Sheol."

More sensible city leaders knew that catching up to New York

City or Chicago was a fanciful ambition. Even Detroit and Cleveland had surged past Buffalo in population. Yet they dreamed of greatness, particularly in midcentury, when the booming economy helped Buffalo become a center for modern art, contemporary music, and, thanks to the new state university, cutting-edge humanities. An advertisement produced by the local Chamber of Commerce in 1963 echoed Pan-American boosters. Buffalo was the "port for 5,000 visiting ships of every flag . . . blast furnace for 7,220,500 tons of steel . . . terminal for more than 20,158,555 tons of rail freight . . . manufacturer of products worth $2 billion." It was "flour mill for the nation" and the "research center of the world."[21]

And then there was a fall—a fall so protracted and deep that it made McKinley's death feel like a bump in the road.

Buffalo, once the grand way station—the famous inland port with grain elevators as imposing as office buildings—became known as a city of rust and struggle. In 1959, a project that had gestated since the 1920s became reality, as engineers completed the Saint Lawrence Seaway, a massive canal that offered shipping companies an opportunity to bypass Buffalo and send their freighters directly from the western Great Lakes through to the Atlantic Ocean.

Thankfully, there was still the power of steel. In 1965, Buffalo produced more than seven million tons of steel at Bethlehem's Lackawanna plant, and, nearby, factories stamped cars out of the metal.

And then that, too, was gone. Bethlehem Steel, succumbing to the triple challenges of foreign competition, new regulations, and management miscalls, first laid off workers and then, in the late 1970s, fired them. Auto plants closed. Manufacturing jobs withered and died. By the early 1980s, big-muscled Buffalo had atrophied. In April 1983, a Buffalo company advertised for forty workers and ten thousand candidates stood in line. Four months later, the Lackawanna steel mill saw its last day of production. Its workers did not go out without comment, however. In the middle of the night on August 16, they raised the international distress signal on top of a blast furnace. The upside-down American flag, measuring more than five hundred square feet and illuminated by two big mercury vapor lamps, blew its message of defiance out over the lake. It was said it could be seen as far north as Niagara Falls.[22]

It was in the wake of these losses, and in the face of a diminishing population—Buffalo went below three hundred thousand in 2000—that the city celebrated the one-hundredth anniversary of the Pan-American Exposition. The university, the historical society, museum curators, city officials, local historians, and enthusiasts of all sorts collaborated to consider the meaning of Rainbow City. They pondered the implications of a fair that, in 1901, had encouraged an interconnected hemisphere. What did the fair now mean in the "global" twenty-first century? How did questions that the fair exposed—about racial justice, economic equality, the American mission, the promise and perils of technology—persist and change? The anniversary inspired a Pan-American Exposition website—with virtual visits to the fair—symposia, exhibits, tours, histories, not to mention a best-selling novel, *City of Light*, which brought to life some of Buffalo's best-known characters.

The anniversary, like the Exposition it honored, generated differences of opinion. Did 1901 represent Buffalo's "shining moment," as some suggested, or did this sort of thinking dismiss the city's promise and future? The *New York Times* weighed in on the debate. In an essay by Randal Archibold announcing the centennial, a headline said it all: "Buffalo Gazes Back to a Time When Fortune Shone: Much-Maligned City Celebrates the Glory of a Century Ago."

The writer began with harsh words: "It is hard for outsiders to imagine that before the steel mills died, the young people fled, and the Bills choked, Buffalo was unbowed and proud." He went on: Buffalonians were indulging in "a spate of nostalgia that has become something of a civic obsession."

If Buffalo was maligned, New York City had, for more than a century, carried some of the guilt. Mismatched as they were, Buffalo and New York had been urban siblings. When opposition came from outside the state in 1901, Manhattan became a fierce protector: helping Buffalo secure the Exposition, helping defend Buffalo against slander after the assassination, and helping Buffalo get funds from the federal government. But, siblinglike, New York City threw punches. In 1901, New Yorkers were accused of seeing the Lake Erie city as a "pretentious little sister" and of not bothering to go to the Exposition "in sufficient numbers."

Now, some one hundred years later, the belittlement seemed to persist. It wasn't just that New Yorkers saw Buffalo as a city of past, not future, magnificence, but a Manhattan urban planner suggested in 2010 that it was perhaps time for certain upstate cities to die. Across the Midwest, Mitchell Moss said, state governments had let once-proud cities disappear, and it was time for New York to "do the same." He chose Buffalo as his example of a city that once flourished, and he used 1901 as his frame of reference.

Buffalo, of course, was not giving up and not going anywhere. It had not given up when it was slammed with an assassination and an insolvent Exposition. It focused on what it had gained and what it had imparted. It had not given up when merchant ships went elsewhere and steel mills failed. It moved forward, reused, and reinhabited. Instead of taking iron ore and making steel, the city began to work at the earth a different way. Along the lake, using roads worn down by steelworkers and grounds flattened by blast furnaces, developers put up wind machines, sentinels of a new age. They planned factories for solar machines. Visionaries, too, reawakened neighborhoods and pulled people back to the core of the city. They turned industrial landscapes into byways and parks and defiantly, boldly, lit grain elevators with flashing colors.

The pride that launched the Exposition in 1901 gave birth to the pride that ushered in Buffalo's "new beginning" in the twenty-first century. Rainbow City itself, however, mostly disappeared. Buffalo honored McKinley with a ninety-six-foot obelisk, funded by the state, that went up in front of City Hall in 1907. It saw a downtown office building, reminiscent of the Electric Tower, rise in 1912, and it handed the house where a tired Theodore Roosevelt took the oath of office to the National Park Service in 1969. With the exception of the New York State Building, however, the 1901 grounds were let go, and, with time, developers seeded them into cropped lawns, put up well-appointed houses, and paved new avenues. Faded plaques mark the locations of some Exposition buildings, and metal tablets tell pedestrians where the president was shot and died, but mostly it is hard to know that, long ago, there stood on the site an enchanted metropolis.[23]

Other evidence of Buffalo's fair, on the other hand, lives on. Remnants of Rainbow City, and accounts of those who animated it, sit in museum exhibits and rest on shelves of local libraries. Records also exist, ever so compactly, in virtual archives. Tireless recordkeepers unlock them for us, not only bringing the Exposition back to life but also allowing it to change over time, with new ways of seeing.

VIII
THE TIMEKEEPERS

The spectacle called Rainbow City had been built by people motivated by a love for their hometown and country and, of course, a desire to make money. They were exceptionally proud of their way of life and believed it should be shown off and shared. In their triumphant march to the apex of civilization, they said, they had not only overcome human savagery but also tamed the animal kingdom and, for the sake of modern technology, conquered the natural world.

Rainbow City was not built without opposition, and, particularly in the autumn of 1901, it did not carry on without resistance from performers and members of the public. Animals acted out, too, and even "nature," like Niagara Falls, and forces, like electricity, took prisoners. The Buffalo Exposition generated modern concerns about the meaning of technology and generated modern discussions about animal welfare. It also spurred talk of social equality and marked the sudden appearance of a modern American president. The Pan-American fair brought the country into the twentieth century with a literal jolt.

There had been other magical exposition cities in the late nineteenth century, and there were more to come. Many echoed the same themes of civilization, nationhood, and globalized trade that had inspired the Pan-American Exposition. All together, they instructed and entertained, and, yes, provoked millions and millions of people.

By the end of the twentieth century, these massive fairs not only dropped in frequency but also shifted shape. While there was little change in the business aims of fairs, and, behind the scenes, corporate capital held its ground, swings in the social tide meant new sorts of displays. Ongoing struggles against imperialism, the civil rights movement, and other efforts for social justice meant a gradual end to displaying "types" of people. Celebrating the subjugation of animals and the planet also lost favor.

Expo '74, a world's fair in Spokane, Washington, reflected this fluctuating landscape. At first glance, the Spokane exposition sounded oddly like the Pan-American fair: a city host eager for development and investors, a waterfall as a central feature, and an exposition without a women's pavilion. But it went in a new direction. Not only did it feature a grand African American pavilion, it also hosted a celebration of the state's recent passage of the Equal Rights Amendment for women. Most striking was its focus on the environment. While it was accused of being a sellout to commercialism, and it lacked the backing of major environmental groups, it hosted serious conferences and exhibits focused on environmental concerns. The United States Pavilion even inscribed on its central wall a bold new message: THE EARTH DOES NOT BELONG TO MAN. MAN BELONGS TO THE EARTH.[24]

The Spokane fair, and others that succeeded it, showed us how things had changed since 1901. These expositions were, as William McKinley claimed just before he was killed, the timekeepers of progress. McKinley's idea of progress—the triumph of big business, the success of the white man's civilizing mission, the harnessing of the globe—was far removed from the meaning of progress that would prevail for subsequent generations.

But McKinley was right: These fantasylands were telling. Rainbow City, intended as a scripted and neatly schemed production, became an improvised performance—where the rich and the powerful, the poor and the desperate, the human and the animal, and the natural world, in all its beauty and fury, met in dynamic alchemy.

It was the supreme measure of a moment in time.

Acknowledgments

It is a pleasure to thank the many people who have made this book possible. At the heart of this project, of course, is the proud and resilient City of Buffalo, with its archives and local historians, and its friends and families.

The staff at the Buffalo and Erie County Public Library first introduced me to the Pan American Scrapbooks (before they were digitized) and to all the collections of the Grosvenor Room. Thank you, Charles Alaimo and Carol Pijacki, for your cheerful and patient support. At the Buffalo History Museum, which, as the former New York State Building, is a memorial to the Pan-American Exposition, Cynthia Van Ness reigns as Queen of the Archives, and she is as astute and knowledgeable as she is tirelessly helpful.

Independent scholar Susan Eck, with her remarkable website, has been hugely generous with her knowledge of all things Pan-American. From traipsing around the city looking for signs, to digging up old maps, to forwarding new materials, she has offered wonderful assistance. She deserves a gift basket as big as a house.

Mark Goldman, "Mr. Buffalo," deserves special credit for not only introducing me to Jumbo II and his almost-demise but also for reintroducing me to the history of the city. Buffalo's vibrancy in the twenty-first century owes a lot to Mark's passion and commitment.

I thank Professor Michael Frisch, too, for insightful conversation and guidance and insight into the meaning of the Pan-American for Buffalo today.

For help with the history of Annie Taylor, I appreciate the good work and help of Dwight Whalen and the inspiration of Monica Wood.

For their assistance and advice, I am also grateful to Melissa Brown, Sandy Starks, Stanton Hudson, Steve Bell, Mary Rech Rockwell, and Brent Baird. Leslie Zemsky, one of the brightest lights of contemporary Buffalo, deserves buckets of roses.

And then there are the Gurneys: Susan and Nancy—the sisters I never had, who have offered me a home away from home, wherever they are, for seemingly forever, and who have been my most loyal cheerleaders. Jackie and Bill Gurney and Elizabeth and Sam Gurney are old friends whose kindness knows no bounds. Elizabeth provided introductions that only a dedicated Buffalonian, with a deep pride in her city, could offer.

At the other end of New York State, in Manhattan, two important people pushed this project to the finish line: Literary agent Jennifer Lyons deftly shepherded the work to its resting place with W. W. Norton, and senior editor Amy Cherry, with her seasoned eye and incisive pen (not to mention exuberant handwriting), improved the narrative tremendously. Any infelicities that exist are solely attributable to the stubbornness of the author. Working with Norton's Remy Cawley was also a distinct pleasure, and copyeditor Kathleen Brandes deserves heartfelt applause for her astute and painstaking work.

Closer to home, Bates College generously granted me research support and travel funds to Buffalo and the United Kingdom, and departmental colleagues provided encouragement and cheer. Will Ash needs a medal for his patience, not to mention his skilled work with images. Students Madeleine McCabe, Hallie Posner, Hannah Gottlieb, Alicia Fannon, Anna Whetzle, and especially the traveling Rebecca Merten deserve many thanks. I am also grateful to my hometown public libraries in Yarmouth and Freeport, Maine, which provided me with soft chairs, hot tea, and, best of all, quiet.

Finally, my family. My ninety-two-year-old mother, the witty and clever author J. S. Borthwick, provided line-by-line editorial advice. Asked how they would like to be acknowledged, my children offered the following: My son Nick wants to be recognized as intelligent and loving and the favorite son. My son Malcolm wants to be recognized as intelligent and loving and the favorite son. My daughter, Louisa, would like to be honored as the favorite child. My daughter-in-law, Julie, too modest to write her own credits, merits recognition as one of the warmest and most spirited people I have ever known.

My husband, Rob Smith, provided life support on this project from day one. From New York to Georgia to Michigan to London, he helped me sleuth out stories of Pan-American characters, and he read drafts with good humor and insight. His steady encouragement, not to mention patience with curmudgeonly moods, deserves my deep and heartfelt gratitude.

Illustration Credits

Pages x–xi: Rainbow City at night. Source: *The Latest and Best Views of the Pan-American Exposition* (Buffalo: Robert Allan Reid, 1901). Photo courtesy of Susan Eck.

Page 13: Buffalo River with grain elevator and freighters. Collection of the Buffalo History Museum, used by permission.

Page 14: Lafayette Square. Source: *One Hundred Views of the Pan-American Exposition, Buffalo, and Niagara Falls* (Buffalo: Robert A. Reid, 1901).

Page 23: Map of the Pan-American Grounds. Image courtesy of Susan Eck.

Page 26: The East Esplanade at night. Collection of the Buffalo History Museum, used by permission.

Page 28: The North Midway. Source: C. D. Arnold, *Official Views of Pan-American Exposition* (Buffalo, 1901).

Page 31: Frank Charles Bostock. Source: Postcard advertisement, n.d. Private collection.

Page 34: Animals line up in front of Bostock's arena. Source: C. D. Arnold, *The Pan-American Exposition Illustrated* (Buffalo, 1901).

Page 37: Chiquita. Source: Elmer Chickering Collection, Houghton Library, Harvard University.

Page 39: Bostock's Main Attractions. Source: *One Hundred Views of the Pan-American Exposition* (Buffalo: Robert A. Reid, 1901).

Page 46: The Indian Congress. Source: *Photographs of the Pan American Exposition Held in Buffalo May 1, 1901 to Nov. 1, 1901* (Charles H. Williams viewbook). Reproduction by Permission of the Buffalo & Erie County Public Library, Buffalo, New York.

Page 48: Jumbo II arrives in Buffalo. Source: *Buffalo Courier*, August 4, 1901. Reproduction by Permission of the Buffalo & Erie County Public Library, Buffalo, New York.

Page 51: Pan-American Power Brokers. Sources (left to right): Courtesy of Susan Eck; Richard H. Barry, *The True Story of the Assassination of President McKinley at Buffalo* (Buffalo: Robert Allan Reid, 1901); *One Hundred Views of the Pan-American Exposition, Buffalo, and Niagara Falls* (Buffalo: Robert A. Reid, 1901).

Page 55: William and Ida McKinley. Collection of the Buffalo History Museum, used by permission.

Page 62: Frank Bostock's invitation to President McKinley. Source: Richard H. Barry, *Snap Shots on the Midway of the Pan-Am Expo* (Buffalo: Robert Allan Reid, 1901).

Page 67: Leon Czolgosz, alias Fred Nieman. Collection of the Buffalo History Museum, used by permission.

Page 78: The Triumphal Bridge on Flag Day. Photographer unknown. Photo courtesy of Susan Eck.

Page 82: Stereographic view of crowds on President's Day. Kilburn, B. W., photographer. "Crowds at the Temple of Music where President McKinley was assassinated, Pan American Exposition." Stereograph. Littleton, NH: B. W. Kilburn, c. 1901. From Library of Congress Prints and Photographs Division, http://www.loc.gov/pictures/resource/cph.3c17986/ (accessed February 25, 2016).

Page 84: President McKinley addresses the crowd, September 5. Collection of the Buffalo History Museum, used by permission.

Page 88: Jim Parker. Source: Richard H. Barry, *The True Story of the Assassination of President McKinley at Buffalo* (Buffalo: Robert Allan Reid, 1901).

Page 90 (left): Martha Wagenfuhrer. Courtesy of the Niagara Falls (NY) Public Library; (right): Section of the Niagara Escarpment. Rendering by Margaret Creighton and William Ash.

Page 98: Fairgoers wait near the Exposition Hospital. Source: Richard H. Barry, *The True Story of the Assassination of President McKinley at Buffalo* (Buffalo: Robert Allan Reid, 1901).

Page 104: Secretary George Cortelyou delivers medical bulletins. Source: Richard H. Barry, *The True Story of the Assassination of President McKinley at Buffalo* (Buffalo: Robert Allan Reid, 1901).

Page 110: Carlisle Graham and Maud Willard. Courtesy of the Niagara Falls (NY) Public Library.

Page 121: Mary Talbert. Collection of the Buffalo History Museum, used by permission.

Page 124 (left): On stage at the Old Plantation. Source: *Photographs of the Pan American Exposition Held in Buffalo May 1, 1901 to Nov. 1, 1901* (Charles H. Williams viewbook). Reproduction by Permission of the Buffalo & Erie County Public Library, Buffalo, New York; (right): "Laughing" Ben Ellington with fellow performers. Source: Richard H. Barry, *Snap Shots on the Midway of the Pan-Am Expo* (Buffalo: Robert Allan Reid, 1901).

Page 131: McKinley's recovery ribbon. Image courtesy of Heritage Auctions.

Page 138: Roosevelt talks with reporters. Collection of the Buffalo History Museum, used by permission.

Page 141: McKinley lies in state. Source: Richard H. Barry, *The True Story of the Assassination of President McKinley at Buffalo* (Buffalo: Robert Allan Reid, 1901).

Page 153: Geronimo and Wenona. Author's collection.

Page 154: Fairgoers peer under the flap of a tepee. Collection of the Buffalo History Museum, used by permission.

Page 158: Lion wedding. Collection of the Buffalo History Museum, used by permission.

Page 165: Annie Taylor, barrel, and cat. Courtesy of the Niagara Falls (NY) Public Library.

Page 172: Mabel Barnes scrapbook excerpt. From Barnes, "Peeps at the Pan-American: an account of personal visits in the summer of 1901 from notes jotted down on the spot and put in permanent form during fourteen years." Mss. W-119. Collection of the Buffalo History Museum, used by permission.

Page 184: Taylor inside barrel. Courtesy of the Niagara Falls (NY) Public Library.

Page 189: Taylor, dazed and triumphant. Courtesy of the Niagara Falls (NY) Public Library.

Page 195: Sham battle. Source: C. D. Arnold, *The Pan-American Exposition Illustrated* (Buffalo, 1901).

Page 204: Chiquita and Bostock. Source: Elmer Chickering Collection, Houghton Library, Harvard University.

Page 207: Darkest Africa. Source: C. D. Arnold, *The Pan-American Exposition Illustrated* (Buffalo, 1901).

Page 226: Mary Lord. Collection of the Buffalo History Museum, used by permission.

Page 240: Big Liz and baby. Source: *Buffalo Times*, September 29, 1901. Reproduction by Permission of the Buffalo & Erie County Public Library, Buffalo, New York

Page 242: Jack Bonavita and lions. "Bostock's Trained Lions." Hall Photograph (NY), 1903. From Library of Congress Prints and Photographs Division, https://www.loc.gov/item/2012645716/ (accessed February 26, 2016).

Page 245: Tony and Alice Woeckener (Chiquita). Author's collection.

Page 250: Annie Taylor sells souvenirs. Courtesy of the Niagara Falls (NY) Public Library.

Page 258: Triumphal Bridge in ruins, June 1903. Collection of the Buffalo History Museum, used by permission.

Notes

NOTE ON SOURCES

This account is drawn from memoirs, legal and medical records, and scholarly literature. To a large extent, though, it is a story grounded in press accounts, and, as everybody knows, newspapers provide a good, but imperfect, window into the past. The big Buffalo papers saw eye-to-eye with business interests in the city, for instance, even while they differed politically, and they did not offer much detail on the concerns of working-class, African American, or immigrant communities. They also differed in their approach to accuracy. While a few Buffalo newspapers—especially the *Buffalo Morning Express*—took pride in "truth and accuracy," others proudly printed embellishments. To the degree it is possible, this account favored the papers that applauded careful reporting, and, in other sources, looked for verification elsewhere. The truth of bygone days, though, as any historian knows, is a slippery thing.

Even as they provided somewhat obstructed views into historical events, the publishers, editors, and reporters described here deserve enormous credit. Not only could some journalists craft the most artful sentences imaginable; they also evoked scenes in ways both compelling and persuasive. On occasion, they also championed the underdog. The papers here were big boosters of the Pan-American Exposition and served as proud spokesmen for its backers and its ideals. Yet they knew a good story when they saw it. To the degree that this narrative discusses the interplay of power, then, and forecasts the social battles of the upcoming century, we must be grateful to the tireless, talented newsmen.

NEWSPAPER ABBREVIATIONS USED IN NOTES
Buffalo Commercial: Com
Buffalo Courier: Courier
Buffalo Enquirer: Enq
Buffalo Evening News: News
Buffalo Morning Express: Express
Buffalo Times: Times

PROLOGUE

1   **Jumbo in the stadium**: Mark Goldman, *High Hopes: The Rise and Decline of Buffalo, New York* (Albany: State University of New York Press, 1983), 18–19; *Courier*, Nov. 10, 1901; *Illustrated Buffalo Express*, Nov. 10, 1901; *Newark Advocate*, Nov. 9, 1901; *Charlotte Daily Observer*, Nov. 12, 1901. **Electricity and civilization**: William S. Aldrich, "Mechanical and Electrical Features of the Pan-American Exposition," *Engineering Magazine* 21 (April–September 1901): 842; Jürgen Martschukat, "'The Art of Killing by Electricity': The Sublime and the Electric Chair," *Journal of American History* 89, no. 3 (December 2002): 901. **Animal King**: Richard H. Barry, *Snap Shots on the Midway of the Pan-Am Expo* (Buffalo: Robert Allan Reid, 1901), 111. **Latin America in expositions**: Lisa Munro, "Investigating World's Fairs: An Historiography," *Studies in Latin American Popular Culture* 28 (2010): 80–94. **Color scheme**: C. Y. Turner, "The Pan-American Color Scheme," *The Independent* 53 (April 25, 1901): 948–49. **The vanishing world**: *Times*, Sept. 8, 1901; John Grant and Ray Jones, *Niagara Falls: An Intimate Portrait* (Guilford, CT: Globe Pequot, 2006), 87.

2   **The grand age of fairs**: Robert W. Rydell, *All the World's a Fair* (Chicago: University of Chicago Press, 1984), 2–3. With his groundbreaking work, Rydell launched a generation of scholarly work on these remarkable (and remarkably common) events. For excellent overviews of fair historiography and methodology, see James Gilbert, *Whose Fair? Experience, Memory, and the History of the Great St. Louis Exposition* (Chicago: University of Chicago Press, 2009), esp. 1–36, 53–68; and Lisa Munro, "Investigating World's Fairs: An Historiography," *Studies in Latin American Popular Culture* 28 (2010): 80–94. **New pocket money**: *Wisconsin Weekly Advocate*, May 16, 1901. **Convenient stopping-off place**: *Express*, Jan. 23, 1899. **Early popularity of fair**: *Enq*, June 27, 1901;

Thomas Leary and Elizabeth Sholes, *Buffalo's Pan-American Exposition* (Charleston, SC: Arcadia Publishing, 1998), 77. **Critics**: See, for example, comments made (in a generally favorable review) by William H. Hotchkiss, "The Pan-American on Dedication Day," in *The American Monthly Review of Reviews* 23 (June 1901): 679; and Robert Grant, "Notes on the Pan-American Exposition," *Cosmopolitan* (September 1901), http://panam1901.org/documents/cosmoarticle.html.

CHAPTER 1: RAINBOW CITY

1   **The White City**: See Erik Larson's masterful narrative, *The Devil in the White City: Murder, Magic, and Madness at the Fair That Changed America* (New York: Crown Publishers, 2003).

2   **Chicago's numbers**: A thorough discussion of visits versus numbers of visitors can be found in Gilbert, *Whose Fair?*, 14–16; **Omaha exposition**: W. A. Rodgers, "The Exposition at Omaha," *Harper's Weekly*, Oct. 8, 1898, quoted in David J. Peavler, "African Americans in Omaha and the 1898 Trans-Mississippi and International Exposition," *Journal of African American History* 93 (Summer 2008): 337; Rydell, *All the World's a Fair*, 124.

3   **Pan American fundraising and banquets**: *Express*, Jan. 23, 25–29, 31, 1899; *Com*, Jan. 27, 1899; Frank Baird, unpublished memoir, private collection.

4   **Trip to Washington**: *Express*, Jan. 30, 1899. **Pan-American Themes**: *Pan-American Exposition Buffalo: Its Purpose and Plan* (Buffalo: Pan-American Exposition Company, 1901), 6, http://digital.hagley.org/cdm/ref/collection/p268001coll12/id/6749, accessed Oct 1, 2015; **United States as comrade and friend**: Rydell, *All the World's a Fair*, 128.

5   **Buffalo's assets and achievements**: *Express*, Jan. 31, 1899; *Com*, Feb. 1, 1899; Oct. 6, 1900; *Express*, Feb. 2, 1899; May 5, 1901; *Pan-American Exposition Buffalo: Its Purpose and Plan*, 6; Samuel G. Blythe, "Buffalo and her Pan-American Exposition," *Cosmopolitan* 29 (May–October 1900): 507–12; Goldman, *High Hopes*, ch. 3, 6.

6   **Sinister signs**: *News*, April 17, 21, 1901. **Albany helping**: William I. Buchanan, *Pan American Exposition: Report of William I. Buchanan, Director-General* (Buffalo, 1902), 54, Special Collections, Buffalo and Erie County Public Library. Albany ultimately allotted $100,000 for the New York State Building, while the City of Buffalo and the Historical Society supported the permanent construction with $25,000 each. Thanks to Susan Eck for this information.

7   **Dedication Day**: *Express*, May 19, 1901; *Com*, May 20, 1901; *News*, May 23, 1901.

8   **Optimism**: *Enq*, June 27, 1901; *News*, May 5, 1901; *Wisconsin Weekly Advocate*, May 16, 1901. **Excursionists**: *Com*, July 9, 1901. **Mexico**: *Courier*, May 19, 1901. **Canada**: *Express*, May 4, 1901. Toronto also put on a popular provincial exposition every fall, and that may have deflected some interest away from Buffalo. See Keith Walden, *Becoming Modern in Toronto: The Industrial Exhibition and the Shaping of Late Victorian Culture* (Toronto: University of Toronto Press, 1997), 20. **Tesla and Edison**: *Express*, March 23, 1901; *Courier*, March 22, July 21, 1901; *Times,* Oct. 20, 1901; *Western Electrician* 29 (August 1901): 103, accessed at http://library.buffalo.edu/pan-am/exposition/electricity/development/edisonatexpo.html.

9   **Cleveland on the costs of the fair**: *Cleveland Plain Dealer*, June 13, 1901. **Leon Czolgosz (Fred Nieman)**: L. Vernon Briggs, *The Manner of Man That Kills: Spencer—Czolgosz—Richeson* (Boston: Richard G. Badger, 1921), 275; A. B. Spurney to H. C. Eyman, Feb. 16, 1902, Dr. Walter Channing Papers, Massachusetts Historical Society; *Com*, Sept. 7, 8, 1901; *News*, Sept. 8, 1901; *Express,* Sept. 9, 1901; *Courier*, Sept. 25, 1901. **On Czolgosz's background and murderous act**, see also the excellent analysis by Eric Rauchway: *Murdering McKinley: The Making of Theodore Roosevelt's America* (New York: Hill and Wang, 2003).

10   **Mabel Barnes**: Mabel E. Barnes, "Peeps at the Pan-American. An account of Personal Visits in the Summer of 1901 from Notes jotted down on the Spot and put in Permanent Form during Fourteen Years," Vols. I–III, handwritten scrapbook, Buffalo History Museum Archives, Mss. W-119. Barnes, who made $600 in annual salary in 1899, lived at 64 Johnson Park and taught second grade at the East Delevan Avenue School. See *Annual Report of the Superintendent of Education 1898–1899* (Buffalo: The Wenborne-Sumner Co. Printers, 1900.) The identity of Barnes's companion Abby Hale, described as "Miss Hale" in the scrapbooks, is safe conjecture. Mabel Barnes lived as a boarder under Abby Hale (who was twenty-seven years older than Barnes) in 1900. They lived together for much of their adult lives, with Barnes assuming the "head of household" position as Hale entered her eighties. See US Federal Census reports for 1900, 1910, 1930, National Archives and Records Administration, accessed at http://home.ancestry.com/.

11   **The colors**: Turner, "The Pan-American Color Scheme": 948–49; Katherine V. McHenry, "Color Scheme at the Pan-American," *Brush and Pencil* 8 (June 1901): 151–56, accessed at http://www.jstor.org/stable/25505650.

12   **The Electric Tower and the Goddess of Light**: Isabel Vaughan James, "The Pan-American Exposition," *Adventures in Western New York*

History 6 (1961): 3, accessed Sept. 30 2015, at http://bechsed.nylearns .org/pdf/The_Pan_American_Exposition.pdf. **Tower and manhood**: Rydell, *All the World's a Fair,* 134–36. **Mabel at the fair**: "Peeps," Vol. II, pp. 65, 82.

13 **The Illumination**: "Peeps," Vol. II, pp. 142–49; Vaughan James, "Pan-American Exposition," 10; *Express*, May 5, 1901; *Com,* Aug. 17, 1901.

14 **The Midway**: *News,* Mar. 26, 1899; Barry, *Snap Shots on the Midway of the Pan-Am Expo,* 1, 19–20, 34; Barnes, "Peeps," Vol. II, pp. 171, 196. **The Midway at night**: *New York Times,* June 16, 1901.

15 **What Mabel missed**: *"Lillian Smith: The On-Target 'California Girl,'"* in http://www.historynet.com/lillian-smith-the-on-target-california -girl.htm, accessed June 9, 2015.

CHAPTER 2: SUMMER IN THE CITY

1 **Bostock's physique**: *The World's Fair,* Oct. 12, 1912, in http://www .nfa.dept.shef.ac.uk/jungle/index1a4.html. **Bostock's application**: *Enq*, March 16, 1899; *News*, March 16, 1899; *Express*, March 17, 1899.

2 **Bostock and wife**: *Sheffield Evening Telegraph*, March 4, 1893; *Illustrated* [London] *Police News,* March 11, 1893; *Dundee* [Scotland] *Courier & Argus*, March 10, 1893, British Library microfilm, accessed April 14, 2012.

3 **Lion School**: *Express*, April 1, 1900. **Daniel in the Den**: *Enq,* July 27, 1900.

4 **Bostock's boasts**: "Frank C. Bostock's Grand Zoological Congress and Trained Animal Arena" (Buffalo: Courier Co., 1901). **Bostock's animals and race**: *Enq*, March 16, 1899; *Courier*, Oct. 22, 1901. **Captain Maitland**: *Courier*, Oct. 18, 1901. **Bostock's bodyguard**: *Times*, July 7, 1901.

5 **Weeden**: *Courier*, June 28, 1901. **Tony and Chiquita**: *Express*, Nov. 9, 1901.

6 **Chiquita's birth**: *Boston Daily Globe,* Dec. 13, 1896. **Career as performer**: *Erie* [PA] *Morning Dispatch,* Sept. 29, 1902; Frank C. Bostock v. Espiridiona Alice Cenda Woeckener, Equity Case No. 27; November Term, 1902, Circuit Court of the United States for the Western District of Pennsylvania, Philadelphia Archives.

7 **Extraordinary bodies**: Rosemarie Garland Thomson, *Extraordinary Bodies: Figuring Physical Disability in American Culture and Literature* (New York: Columbia University Press, 1997), 55–70. **Chiquita as mascot**: *Com*, July 10, 1901. **Imagining Cuba**: Louis A. Perez, Jr., *Cuba in the American Imagination: Metaphor and the Imperial Ethos* (Chapel Hill: University of North Carolina Press, 2008), 71–117. **Chiquita and McKin-**

ley: *New York Times*, Feb. 14, 1901. **Not a doll**: *Boston Globe,* Dec. 13, 1896.

8   **Chiquita and Tony**: *Erie* [PA] *Morning Dispatch,* Nov. 4, 1901; March 8, 1902; *Express,* Nov. 9, 1901.

9   **Feeling the fair**: Diary of Levant F. Hillman, Jan. 4–Nov. 3, 1901, unpublished manuscript, Buffalo History Museum Archives, Mss. A2001–5. **Annie Taylor**: Orrin E. Dunlap, "Interview with Mrs. Taylor," Oct. 25, 1901, unpublished manuscript, Stunts and Stunters file, Local History Collection, Niagara Falls [NY] Public Library, hereafter NFPL; Annie Edson Taylor, *Over the Falls: Annie Edson Taylor's Story of Her Life* (privately printed, 1902), reprint.

10 **Attendance concerns**: *Com*, July 9, 23, 1901; *Courier*, Aug. 2, 1901. **Chicago swelters, Philadelphia languid**: *Express*, July 21, 1901. **Reducing admission price**: *Express,* July 21, 1901. **Utah Day**: *Courier*, July 25, 1901. **Women's exhibits**: *Express*, April 23, 1901; *New York Evening Post*, Nov. 21, 1900; *Enq,* Feb. 2, 1901; Marian DeForest, secretary of the Board of Women Managers, explained that the board wanted to show the work of women "because it was good," not simply because it was produced by females. However, there were enough women who wanted a distinctive venue that a small space in the Manufactures Building was given over to displaying "women's" work. The items on display won numerous awards. See DeForest report, *Express*, Nov. 2, 1901.

11 **European exhibits**: See Moses P. Handy, ed., *The Official Directory of the World's Columbian Exposition, May 1st to October 30th, 1893.* Chicago: W. B. Conkey Co., 1892, accessed May 18, 2015, at https://archive.org/stream/officialdirector00worl/officialdirector00worl_djvu.txt; Buchanan, *Pan American Exposition Report,* 54. **Buenos Aires**: Roy Crandall, "Friendly Cooperation," *Pan American Herald* 1 (December 1899): 3. **Backhanded compliments**: *Times,* Oct. 20, 1901; *Express*, July 21, 1901; *Com,* Aug. 28, 1901. See also Mark Bennitt, *The Pan-American Exposition and How to See It* (Buffalo: The Goff Company, 1901).

12 **Midway Day schemes**: *Courier*, July 23, 1901; *Express*, July 24, 1901.

13 **Midway Day parade**: *Express,* Aug. 3, 4, 1901; *Com,* Aug. 3, 1901.

14 **Brooklyn's Jumbo II**: *Brooklyn Eagle,* May 13, 1900. **Bostock's Jumbo II**: *Courier,* July 24, 31, 1901; *Express,* July 27, 1901; *Com*, July 29, 1901; *News,* Aug. 11, 1901. **Photography float**: *Express,* Aug. 3, 4, 1901. **Mabel Barnes**: "Peeps," Vol. II, p. 152. **Fred Nieman**: Briggs, *The Manner of Man,* 277.

CHAPTER 3: THE FAVORED GUEST

1   **Midway Day attendance**: *Express*, Aug. 4, 1901. **High noon**: *Courier*, Aug. 2, 1901. **Future events**: *Express*, Aug. 18, 1902.

2   **Buchanan**: Rydell, *All the World's a Fair*, 130; *American Journal of International Law* 4 (January 1910): 160–61; *Express*, July 24, 1901. **Milburn**: *New York Times*, June 23, 1901; *San Francisco Call*, Feb. 11, 1900; *New-York Tribune*, June 18, 1901; Susan Eck, "The Milburns and their Famous Home: 1168 Delaware Avenue," http://wnyheritagepress.org/photos_week_2009/mckinley_marker/milburn_house/milburn_house.htm. **Diehl**: *Buffalo Courier Record*, Dec. 20, 1897; *Express*, Jan. 25, 1899. **Diehl and opponents**: *Com*, Feb. 28, 1899; *Express*, March 3, 1899; *New-York Tribune*, Oct. 11, 1897; *Courier*, April 27, 1900.

3   **William McKinley**: *Express*, Sept. 14, 1901. **Grasshoppers**: Howard Wayne Morgan, *William McKinley and His America* (Kent, OH: Kent State University Press), 472. **Emperor**: *Montgomery* [MO] *Tribune*, Aug. 23, 1901. **On McKinley before the fair**: See especially Rauchway, *Murdering McKinley*, 4–8.

4   **Ida McKinley**: Carl Sferrazza Anthony, *Ida McKinley: The Turn of the Century First Lady Through War, Assassination, and Secret Disability* (Kent, OH: Kent State University Press, 2013), ebook location 2885; 4852; John C. DeToledo, et al., "The Epilepsy of First Lady Ida Saxton McKinley," *Southern Medical Journal* 90 (March 2000): 267.

5   **McKinley in Canton**: *Courier*, Sept. 7, 1901. **Presidential security**: Matthew C. Sherman, "Protecting the First Citizen of the Republic: Presidential Security from Thomas Jefferson to Theodore Roosevelt" (PhD Diss., Saint Louis University, 2011), 22–24; 83–84; 96–98; 140–45; 158, 166, 170.

6   **Ida McKinley and George Cortelyou security worries**: Anthony, *Ida McKinley,* ebook, location 3968; 4653; Sherman, "Protecting the First Citizen," 182, 197, 207. **Anarchism in the United States**: Chris Vials, "The Despotism of the Popular: Anarchy and Leon Czolgosz at the Turn of the Century, *Americana: The Journal of American Popular Culture* 3 (Fall 2004), accessed Nov. 30, 2015, at http://www.americanpopularculture.com/journal/articles/fall_2004/vials.htm; Sidney Fine, "Anarchism and the Assassination of McKinley," *American Historical Review* 60 (July 1955): 777–80. **McKinley's confidence**: *Courier*, Sept. 9, 1901.

7   **Arcusa**: *Courier*, Aug. 12, 1901.

8   **Czolgosz remembered**: Briggs, *The Manner of Man*, 277; *Express*, Sept. 9, 1901; *Com*, Sept. 8, 1901.

9   **Bostock's invitation**: *Express*, Aug. 16, 1901. **Bostock's popularity**: *Com*, Aug. 12, 1901. **Tiny Mite**: *Express*, Aug. 10, 13, 1901; *Com*, Aug. 10, 1901.

10   **Ptolemy**: *St. John Daily Sun*, July 13, 1901.

11   **Floodgate**: *Express*, Aug. 26, 1901. **Midway smell**: *Courier*, Aug. 26, 1901.

12   **McKinley plans**: *Express*, Aug. 17, 1901; Sept. 3, 4, 1901; *Com*, Sept. 3, 1901.

13   **Free from serious crime**: *Express*, July 21, 1901. **Bull's warning**: *Annual Report of the Board of Police of the City of Buffalo for the Year Ending December 31, 1901* (Buffalo: Wenborne-Sumner Co., 1902), 22–25. **Criminal list**: *Annual Police Report*, 35–44; *Express*, Oct. 22, 1901.

14   **Nieman applies to the boardinghouse**: Briggs, *The Manner of Man*, 278–79. **Conversations**: *Express*, Sept. 9, 1901; *Courier*, Sept. 9, 1901. **Pumpkin-head**: Briggs, *Manner*, 278.

15   **Czolgosz's illnesses**: Briggs, *Manner*, 293; A. B. Spurney to H. C. Eyman, Feb. 16, 1902, Dr. Walter Channing Papers, Massachusetts Historical Society; Vernon Briggs notes, interview with Emil Schilling, ca. June 1902, in Channing Papers, MHS; Rauchway, *Murdering McKinley*, 118, 167, 204–5. Eric Rauchway suggests that Czolgosz may have been consumed by fears he had syphilis. See Rauchway, *Murdering McKinley*, 180–81.

16   **Nieman/Czolgosz: work, sickness, and disillusionment**: Briggs, *Manner*, 303–8, 314. **Goldman lecture**: *Courier*, Sept. 8, 1901.

17   **Czolgosz and Schilling**: Vernon Briggs notes, interview with Emil Schilling, ca. June 1902, in Walter Channing Papers, Massachusetts Historical Society. **Czolgosz, Goldman, and Isaak**: Abraham Isaak to Walter Channing, June 9, 1902, Channing Papers, MHS; Rauchway, *Murdering McKinley*, 100–4.

18   **Milburn on McKinley's foreign policy**: *Times*, Oct. 30, 1900. **Latin America at world's fairs**: Ines Dussel, "Between Exoticism and Universalism: Educational Sections in Latin American Participation at International Exhibitions, 1860–1900," *Paedagogica Historica* 47 (October 2011): 601, 605–7, 616; Buchanan, *Report of the Director-General*, 26–27; Alvaro Fernandez-Bravo, "Ambivalent Argentina: Nationalism, Exoticism, and Latin Americanism at the 1889 Paris Universal Exposition," *Nepantla: Views from South* 2 (January 2001): 115–39; Nancy Egan, "Exhibiting Indigenous Peoples: Bolivians and the Chicago Fair of 1893," *Studies in Latin American Popular Culture* 28 (January 2010): 7–15.

19   **Chilean minister dies**: *Com*, July 20, 1901; *Express*, August 8, 21, 1901. **Mexico at the fair**: Janice Lee Jayes, *The Illusion of Ignorance: Constructing the American Encounter with Mexico, 1877–1920* (Lanham, MD: University Press of America, 2011), 189; *Com*, Oct. 1, 1901.

20   **Cuba Day remarks**: *Com*, Aug. 29, 1901; *Express*, Aug. 30, 1901.

CHAPTER 4: THE BLOOD-COLORED TEMPLE

1   **Leaving Ohio**: *Marietta Daily Leader*, Sept. 4, 1901; *Saint Paul* [MN] *Globe*, Sept. 4, 1901; [Washington, DC] *Evening Times*, Sept. 4, 1901; *News*, Sept. 4, 1901.

2   **Arrival in Buffalo**: *Com*, Sept. 4, 1901; *Express*, Sept. 5, 1901; Chris Vials, "The Despotism of the Popular: Anarchy and Leon Czolgosz at the Turn of the Century," *Americana: The Journal of American Popular Culture* 3 (Fall 2004), accessed October 8, 2015, at http://www.americanpopularculture.com/journal/articles/fall_2004/vials.htm.

3   **Barbershop**: *Com*, Sept. 9, 1901. **Gun**: *News*, Sept. 8, 1901.

4   **The Triumphal Bridge and government exhibits**: *Express*, Mar. 24, 1901; Vaughan James, "The Pan-American," 8; Mabel Barnes, "Peeps," Vol. I, pp. 27, 105–6. **Nieman blaming McKinley**: Carlos F. Mac-Donald, "The Trial, Execution, Autopsy and Mental Status of Leon F. Czolgosz, Alias Fred Nieman, the Assassin of President McKinley," *American Journal of Insanity* 58 (January 1902): 384, accessed Oct. 10, 2015, at http://mckinleydeath.com/documents/journals/AJI58-3b.htm.

5   **The people's fair**: *Com*, Jan. 24, 25, 30, 1899; *Express*, Jan. 23, 1899; *Enq*, Jan. 24, 1899. **Strikes**: *Times*, Aug. 20, 25, 1900; *Enq*, Sept. 20, 1900; *Express*, Oct. 11, 1900; *Com*, Oct. 13, 1901. **Parade**: *Com*, Sept. 2, 1901. **Gompers**: *Express*, Sept. 3, 1901.

6   **Workers at the fair at dawn, at night**: *Com,* May 20, 1901; *Courier,* July 21, 1901; *Express,* June 29, August 1, 1901. The fair's working men and women are given only slight mention in the newspapers covered by the twenty-four Pan-American Exposition scrapbooks.

7   **Scheme for the poor**: *Courier,* Aug. 17, 1900. **Costs of the fair**: Leary and Sholes, *Buffalo's Pan-American,* 28; *Express,* June 28, August 2, 3, 1901; *Com,* Aug. 3, 1901. **Women's Building**: Report of Marian DeForest, *Express,* Nov. 2, 1901; *News,* June 23, 1901. **Labor Day at the fair for the first time**: *Express,* Sept. 3, 1901.

8   **President's Day**: *Express*, Sept. 5, 6, 1901. **Mabel's twenty-second visit**: Barnes, "Peeps," Vol. III, p. 132.

9   **McKinley's speech, suspicious characters**: *Express*, Sept. 6, 1901. **Nieman/Czolgosz confession**: *Iowa State Register,* Sept. 8, 1901, accessed at http://mckinleydeath.com/documents/newspapers/ISR46-211gp.htm.

10 **Bostock on President's Day**: *Com,* Sept. 6, 1901. **The McKinleys tour and dine**: *Express,* Sept. 6, 1901; *News,* Sept. 6, 1901. **Fireworks**: Barnes, "Peeps," Vol. III, pp. 150–51, 162. **Foreboding**: *News*, Sept. 8, 1901. **Nieman waits**: *News*, Sept. 8, 1901.

11 **James Parker**: *News*, Sept. 8, 1901; *Macon* [GA] *Telegraph* Sept. 9, 1901;

[Omaha] *Morning World Herald,* Sept. 16, 1901. **McKinley and African Americans**: Mitch Kachun, "'Big Jim' Parker and the Assassination of William McKinley: Patriotism, Nativism, Anarchism, and the Struggle for African American Citizenship," *The Journal of the Gilded Age and Progressive Era* 9 (January 2010): 104; Rauchway, *Murdering McKinley,* 71.

12 **Delightful day**: *Courier,* Sept. 9, 1901.

13 **Martha's stunt**: *Niagara Falls* [NY] *Gazette,* Sept. 3, 1926; *Express,* Sept. 7, 1901; Edward T. Williams, "Martha E. Wagenfuhrer, 'Maid of the Rapids,'" unpublished manuscript in Stunts and Stunters file, NFPL; Orrin E. Dunlap, "Martha E. Wagenfuhrer," unpublished typewritten account, Stunts and Stunters file, NFPL. (Martha was also known as *Maggie* Wagenfuhrer.)

14 **McKinley at Niagara Falls**: McKinley: "The President at Niagara," *Street Railway Journal* 18 (Sept. 21, 1901): 330. **Forebodings at lunch**: *Courier,* Sept. 9, 1901. **Nieman at Niagara Falls**: Trial Transcript: "The People of the State of New York against Leon F. Czolgosz." Unpublished trial transcript. 23–24, 26 Sept. 1901, pp. 59–60, accessed Oct. 11, 2015, at http://mckinleydeath.com/documents/govdocs/transcript.htm.

15 **McKinley at the Temple of Music**: *Courier,* Sept. 8, 1901; Trial transcript: "People v. Czolgosz," 12–13.

16 **The reception/attack in the Temple**: *Express,* Sept. 8, 9, 10, 13, 1901; Trial transcript, "People v. Czolgosz," accessed at http://mckinleydeath.com/documents/govdocs/transcript.htm. **Parker**: *News,* Sept. 8, 1901; *Courier,* Sept. 7, 1901; Rauchway, *Murdering McKinley,* 61–65.

CHAPTER 5: THE EMERGENCY

1 **Shooting aftermath**: Trial transcript, "People v. Czolgosz," accessed at http://mckinleydeath.com/documents/govdocs/transcript.htm; *News,* Sept. 8, 1901; *Express,* Sept. 13, 1901; *Courier,* Sept. 7, 1901; DeB. Randolph Keim, "Personal Notes on the Shooting of President McKinley at Buffalo N.Y. Sept. 7, 1901," manuscript transcript, accessed Oct. 10. 2015, at http://www.shapell.org/manuscript/eyewitness-account-of-the-assassination-of-president-mckinley.

2 **Dr. Park in Niagara Falls**: Roswell Park, "Reminiscences of McKinley Week," typed manuscript, Buffalo History Museum Archives, Mss. A00–390. **Drs. Mann and Mynter perform surgery**: Presley M. Rixey, Matthew D. Mann, Herman Mynter, Roswell Park, Eugene Wasdin, Charles McBurney, and Charles G. Stockton. "The Case of President McKinley," *Medical Record* 60 (Oct. 1901) 601–3, accessed Oct. 20, 2015, at http://mckinleydeath.com/documents/journals/MR60-16dp.htm;

"The Case of the Late President McKinley," *British Medical Journal* 2 (November 1901): 1348. **Dr. Park arrives, urges unity**: "Reminiscences," p. 7.

3 **Operating-room issues**: Roswell Park, "Reminiscences"; Rixey et al., "The Case of President McKinley":601–3; "The Case of the Late President," 1348–49; Jack C. Fisher, *Stolen Glory: The McKinley Assassination* (La Jolla, CA: Alamar Books, 2001), 75–81; "The Surgical and Medical Treatment of President McKinley," *Journal of Medicine and Science* 7 (October 1901): 389–90.

4 **The Illumination mistake**: DeB. Randolph Keim, "Personal Notes," p. 7.

5 **Frank Baird**: Recollections, private collection. **Governor**: *Express*, Sept. 7, 1901.

6 **Roosevelt learns of shooting**: *Burlington Free Press*, Sept. 7, 1901; [Brattleboro] *Vermont Phoenix*, Sept. 13, 1901; *St. Albans Daily Messenger*, Sept. 7, 1901; J. B. Burnham, "Vermont League Outing," *Forest and Stream* 57 (September 1901): 208–9.

7 **McKinley at the Milburn house**: *Express*, Sept. 8, 1901; *Com*, Sept. 8, 1901; *Courier*, Sept. 8, 1901.

8 **Reporters**: *Express*, Sept. 9, 13, 1901; *Com*, Sept. 8, 1901; *Courier*, Sept. 8, 9, 1901. **McKinley holds his own**: *Com*, Sept. 8, 1901; *Courier*, Sept. 8, 1901. **McKinley and wife**: *Express*, Sept. 8, 1901.

9 **Nieman/Czolgosz in jail**: *Express*, Sept. 8, 9, 11, 1901; *Com*, Sept. 9, 11, 1901. **Reveals name**: *Express*, Sept. 8, 1901. **On anarchy**: Rauchway, *Murdering McKinley*, 17–19.

10 **Anarchists attacked**: Sidney Fine, "Anarchism and the Assassination of McKinley," *American Historical Review* 60 (July 1955): 785–87; *Com*, Sept. 11, 1901. **Lynch law**: *Com*, Sept. 11, 1901; Vials, *Despotism of the Popular*, p. 8; Kachun, "'Big Jim' Parker," 93–116. **Goldman**: *Com*, Sept. 7, 1901; *Courier*, Sept. 11, 1901. **Socialists**: *Express*, Sept. 9, 1901.

11 **The Exposition in the wake of the shooting**: *Courier*, Sept. 8, 9, 1901; *Com*, Sept. 7, 8, 9, 13, 1901; *Express*, Sept. 9, 1901. **Bostock readjusts**: *Com*, Sept. 8, 1901; *Express*, Sept. 8, 1901.

12 **Maud Willard's fatal trip**: *Express*, Nov. 15, 1901; Edward T. Williams, "Maud Willard Meets Death in Whirlpool," in Stunts and Stunters file, Maud Willard folder, NFPL; Orrin E. Dunlap, "Maud Willard," typed account, Willard file, NFPL; *Niagara Falls* [NY] *Journal*, Sept. 13, 1901; *Niagara Falls* [NY] *Review*, Aug. 23, 1993.

13 **Annie Taylor**: Orrin E. Dunlap, "Interview with Mrs. Taylor, October 25, 1901," unpublished typed manuscript, Stunts and Stunters file, NFPL.

14   **Cautious optimism**: *Com,* Sept. 8, 1901; *Express,* Sept. 10, 1901. **More certain relief**: *Express,* Sept. 9, 10, 11, 12, 1901; *Com,* Sept. 8, 9, 10, 1901; *Courier,* Sept. 8, 1901.

15   **Comparison with Garfield**: *Courier,* Sept. 8, 1901; *Express,* Sept. 10, 1901. **Senator Hanna bubbles over**: *Express,* Sept. 10, 1901. **Mrs. McKinley**: *Com,* Sept. 12, 1901; Carl Sferrazza Anthony, *Ida McKinley: The Turn of the Century First Lady Through War, Assassination, and Secret Disability* (Kent, OH: Kent State University Press, 2013), ebook, location 5128 (chapter 16).

16   **Suffragists visit**: *Express,* Sept. 8, 9, 1901. **Roosevelt on suffrage, 1898, and women's duty, 1905**: "Woman's Column" XI (January 1898), accessed Oct. 14, 2015, at https://archive.org/stream/WomansColumn18981899/ Womans%20Column%201899_djvu.txt; *Address by President Roosevelt before the National Congress of Mothers,* March 2, 1905. Theodore Roosevelt Collection. MS Am 1541 (315), Harvard College Library. http:// www.theodorerooseveltcenter.org/Research/Digital-Library/Record .aspx?libID=o280100. Theodore Roosevelt Digital Library, Dickinson State University. More than a decade later, as a third-party candidate for president, Roosevelt would support women's right to vote unequivocally. **Roosevelt saunters about Buffalo**: *Courier,* Sept. 9, 10, 1901; *Express,* Sept. 9, 1901. **Roosevelt and McBurney leave**: *Express,* Sept. 11, 1901.

17   **Jim Parker as hero**: *News,* Sept. 8, 1901; *Express,* Sept. 10, 1901; *Com,* Sept. 11, 13, 1901; *Washington Post,* Sept. 10, 1901; *Courier,* Sept. 11, 1901.

18   **African American representation at world's fairs**: Robert W. Rydell, "'Darkest Africa': African Shows at America's World's Fairs, 1893–1940," in Bernth Lindfors, ed., *Africans on Stage: Studies in Ethnological Show Business* (Bloomington: Indiana University Press, 1999), 135–45; Sara S. Cromwell, "Fair Treatment? African-American Presence at International Expositions in the South, 1884–1902" (MA thesis, Wake Forest University, 2010), ch. 3; Amma Y. Ghartey-Tagoe, "The Battle Before the Souls of Black Folk: Black Performance in the 1901 Pan-American Exposition" (PhD diss., NYU, 2009), 15–18; Ida B. Wells, ed., "The Reason Why the Colored American is not in the World's Columbian Exposition," http://digital.library.upenn .edu/women/wells/exposition/exposition.html). **Buffalo protests, Mary Talbert**: *Com,* Nov. 12, 1900; *Times,* Nov. 12, 1900; *Express,* July 8, 1901; William H. Loos, Ami M. Savigny, Robert M. Gurn, and Lillian S. Williams, *The Forgotten "Negro Exhibit": African American Involvement in Buffalo's Pan-American Exposition, 1901* (Buffalo: Buffalo and Erie County Public Library and the Library Foundation of Buffalo

and Erie County, 2001); Peggy Brooks-Bertram and Barbara Seals Nevergold, *Uncrowned Queens: African American Women Community Builders* (Buffalo: Uncrowned Queens Publishing, 2005), 163; and http://www.buffalonian.com/history/articles/1901-50/ucqueens/negro_exhibit_at_pan_am.htm. **Negro Exhibit**: *Illustrated Buffalo Express,* 1901, from http://www.fultonhistory.com/Fulton.html.

19 **African American newspapers on the Negro Exhibit**: See, for example, [Kansas City] *American Citizen,* May 17, 1901. *Colored American,* Aug. 10, 17, 1901. **Pan-American Du Bois exhibit**: *Express,* April 24, May 5, 14, 1901; *New York Times,* Sept. 21, 1901, accessed at http://search.proquest.com/docview/96147725?accountid=8505; Loos et al., "The Forgotten 'Negro Exhibit.'"

20 **African Americans and Africans on the Midway**: *Express,* June 25, 29, 1901. **Esau**: *Courier,* Aug. 8, 1901. **Mabel Barnes in Darkest Africa**: "Peeps," Vol. III, pp. 70–88; Robert Rydell discusses ways in which African performers resisted or turned tables on visitors in Chicago in 1893. See "'Darkest Africa,'" 145.

21 **Laughing Ben**: *Com,* Aug. 8, 1901; *Express,* May 12, 1901.

22 **Redefining, erasing Jim Parker**: *Express,* Sept. 9, 10, 12, 13, 1901; *News,* Sept. 8, 1901; *Com,* Sept. 13, 1901; see also Rauchway, *Murdering McKinley.*

CHAPTER 6: THE RISE AND THE FALL

1 **Buffalo's doctors and residents praised**: *Express,* Sept. 10, 1901; *Courier,* Sept. 9, 1901; *Brooklyn Eagle,* reprinted in *Com,* Sept. 11, 1901. **Buffalo as world's epicenter**: *Courier,* Sept. 9, 10, 1901; *Com,* Sept. 9, 1901; *Express,* Sept. 8, 13, 1901.

2 **Upcoming attractions, Railroad Day**: *Express,* Sept. 9–13, 1901; *Com,* Sept. 9, 1901. **Bostock's new publicity**: *Express,* Sept. 12, 1901; *Com,* Sept. 10, 13, 1901; *Courier,* Sept. 13, 1901.

3 **National Jubilee Day plans**: *Express,* Sept. 11, 12, 1901; *Courier,* Sept. 10, 1901.

4 **The change**: *Express,* Sept. 11, 12, 13, 1901; *Com,* Sept. 10, 12, 13, 1901.

5 **New symptoms; alarm; sending word**: Nelson W. Wilson, "Details of President McKinley's Case," *Buffalo Medical Journal* 57 (October 1901): 216; Rixey et al., "The Official Report on the Case of President McKinley," *Buffalo Medical Journal* 57 (October 1901): 280–83; *Express,* Sept. 13, 1901; *Com,* Sept. 13, 1901. **Kipling**: *Boston Medical and Surgical Journal* 140 (March 1899): 269.

6 **Ida McKinley, the weather**: *Com,* Sept. 13, 1901.

7 **Roosevelt informed**: Jacob A. Riis, *Theodore Roosevelt: The Citi-*

zen (New York: Macmillan, 1912), 242–49, accessed at http://babel
.hathitrust.org/cgi/pt?id=mdp.39015008340195;view=1up;seq=261; *New
York Sun*, Sept. 14, 1901; *Com*, Sept. 13, 1901; *Courier*, Sept. 13, 14, 1901.

8   **Bulletins, premonitions, desperate efforts**: Rixey et al., "Official Report," 280–83; *New York Sun*, Sept. 14, 1901; "The People of
the State of New York against Leon F. Czolgosz." Unpublished trial
transcript, 23–24, 26 Sept. 1901 (testimony of Dr. Mann), accessed at
http://mckinleydeath.com/documents/govdocs/transcript.htm.

9   **Morphine; final words**: Rixey et al., "Official Report," 280–83; *Courier*, Sept. 15, 1901; *Com*, Sept. 13, 1901; Anthony, *Ida McKinley*, ebook,
location 5397, chapter 16.

10   **Superintendent Bull**: *Express*, Sept. 14, 1901; *Com*, Sept. 13, 1901.
Coroner: *Express*, Sept. 14, 1901. **Last minutes of life**: *Daily Alaska
Dispatch*, Sept. 18, 1901; "Those Present at the Death-Bed," *Harper's
Weekly* (September 21, 1901): 946; Wilson, "Details of President McKin-
ley's Case," 207–25; *Florence Times*, Sept. 20, 1901.

11   **Pausch**: John Elfreth Watkins, Jr., "M'Kinley Death Mask," [India-
napolis] *Sunday Journal*, Dec. 29, 1901; *New York Times*, Nov. 19, 1901.
Autopsy: Rixey et al., "Official Report," 284–93; "The People of the
State of New York against Leon F. Czolgosz," Trial Transcript, testimony of Herman Mynter; autopsy report of Dr. Harvey Gaylord.
Thanks to infectious-disease specialist Robert P. Smith, MD, patholo-
gist Frederick Meier, MD, and trauma surgeon Stanley Trooskin, MD,
for contemporary insights into this case.

12   **Roosevelt arrives, takes oath of office**: *Com,* Sept. 14, 1901; *Boston
Globe*, Sept. 15, 1901; Marshall Everett, *Complete Life of William McKinley and Story of his Assassination* (Chicago: C. W. Stanton, 1901), 304–5,
accessed at http://babel.hathitrust.org/cgi/pt?id=uc2.ark:/13960/
t5fb4x97r;view=1up;seq=7. **The Exposition in shock and dark**:
Express, Sept. 15, 1901.

CHAPTER 7: AFTERSHOCK

1   **Exposition in mourning**: *Express*, Sept. 16, 1901. **Cortege**: *Com*,
Sept. 16, 1901; *New York Tribune*, Sept. 16, 1901; *Courier*, Sept. 15, 1901.
City Hall mourners: *New York Tribune*, Sept. 16, 1901; *Express*, Sept.
15, 16, 1901; Marshall Everett, *Complete Life of William McKinley and
Story of his Assassination* (Chicago: C. W. Stanton, 1901), 343–44; Doc
Waddell, manager of the Indian Congress, likely wrote Geronimo's
note. See Kevin D. Shupe, "Geronimo Escapes: Envisioning Indian-
ness in Modern America," PhD diss., George Mason University, 2011.

2   **Funeral train, Washington**: *New York Tribune*, Sept. 17, 1901; *Com*,

Sept. 16, 18, 1901; *Express,* Sept. 18, 1901; Everett, *Complete Life of William McKinley,* 345–48.

3   **Canton**: *Express,* Sept. 17, 19, 20, 1901.

4   **Blaming Buffalo surgeons**: *Express,* Sept. 20, 1901; *New York World,* Sept. 16, 1901; *Courier,* Sept. 18, 1901. **Defending Buffalo surgeons**: *Express,* Sept. 18, 1901; *Com,* Sept. 18, 30, 1901.

5   **Gloom**: *New York Times,* Sept. 23, 1901; *Courier,* Sept. 16, 1901; *Express,* Sept. 17, 1901; *Com,* Sept. 21, 1901; *Times,* Sept. 22, 1901. **McKinley's shrine**: *New York Times,* Sept. 23, 1901; *Courier,* Sept. 23, Oct. 1, 2, 1901; *Express,* Sept. 16, 1901.

6   **Bostock rallies**: *Courier,* Sept. 15, 1901. **New animals**: *Com,* Sept. 16, 18, 20, 21, 1901; *Courier,* Sept. 19, 23, 27, 1901; *Express,* Sept. 16, 1901. **Humane Society report**: *Erie County Society for the Prevention of Cruelty to Animals Annual Report* (Buffalo, 1901), 30–33.

7   **Czolgosz's indictment**: *Express,* Sept. 17, 1901. **Praise for trial**: *Com,* Sept. 18, 1901; *Express,* Sept. 17, 1901; Leroy Parker, "The Trial of the Anarchist Murderer Czolgosz," *Yale Law Journal* 11 (Dec. 1901): 80–94, accessed at http://mckinleydeath.com/documents/journals/YLJ11-2.htm; *Daily Picayune,* Sept. 24, 1901. **The trial and sentencing**: *Com,* Sept. 24, 1901; Parker, "The Trial of Czolgosz"; Carlos F. MacDonald, "The Trial, Execution, Autopsy and Mental Status of Leon F. Czolgosz, Alias Fred Nieman, the Assassin of President McKinley," *American Journal of Insanity* 58 (Jan. 1902): 369–86, accessed at http://mckinleydeath.com/documents/journals/AJI58-3b.htm; "The Trial of Czolgosz," *Outlook* 69 (Oct. 1901): 242–43. **Goldman and alienists**: Emma Goldman, "October Twenty-Ninth, 1901," *Mother Earth* 6 (October 1911): 232–35, accessed at http://mckinleydeath.com/quotes/trial.htm; "The Manner of Man that Kills: A Review," *The Journal of Heredity* 13 (March 1922): 136.

8   **Eliminating Parker**: See Mitch Kachun, "'Big Jim' Parker and the Assassination of William McKinley: Patriotism, Nativism, Anarchism, and the Struggle for African American Citizenship," *The Journal of the Gilded Age and Progressive Era* 9 (January 2010): 99. **Conflicting opinions**: *Omaha Daily Bee,* Oct. 7, 1901; *Express,* Sept. 27, 28, 1901; *Courier,* Sept. 26, 28, 1901; *News,* Sept. 27, 1901, Oct. 1, 1901; Kachun, "'Big Jim' Parker," 99. **Vine Street Church meeting**: *Express,* Sept. 28, 1901. **Parker lectures**: *Washington Times,* Oct. 9, 1901; *Colored American,* Oct. 12, 1901.

9   **Cold weather**: *Courier,* Sept. 15, 21, 24, 27, Oct. 3, 5, 22, 1901; *News,* Sept. 24, 1901; *Express,* Sept. 26, 1901; *Com,* Oct. 7, 1901.

10   **Dog feast numbers**: *Courier,* Oct. 6, 1901; *Com,* Sept. 30, 1901.

Geronimo: *Express*, June 29, Sept. 25, Oct. 5, 1901; *Courier*, June 29, Sept. 25, 1901; *Com*, Oct. 1, 3, 1901. **Indians turn the tables**: *Com*, Sept. 2, 1901; *Enq*, June 22, 1901.

11   **Dog feasts**: *Com*, Sept. 28, 30, 1901. **Native ritual**: *Omaha World Herald*, April 21, 1899, Aug. 21, 1898; *Duluth News Tribune*, Aug. 16, Nov. 19, 1899; *Biloxi* [MS] *Daily Herald*, Oct. 26, 1900. **Taking of Buffalo dogs**: *Express*, Sept. 22, 1901; *Com*, Sept. 24, 1901. **Protests**: *Express*, Sept. 25, 1901; ECSPCA, *Annual Report*, 21–22. **Feast**: *Express*, Sept. 27, 1901; *Com*, Sept. 27, 1901.

12   **More animals**: *Express*, Oct. 6, 1901. **Chiquita and Tony**: [Erie] *Daily Times*, Nov. 2, 1901; *Kalamazoo Gazette*, Nov. 13, 1908; *Express*, Nov. 9, 1901; Al Stencell, "Frank Bostock in America," in National Fairground Archive, The Sheffield University, accessed Aug. 13, 2014, at http://www.nfa.dept.shef.ac.uk/jungle/index1a1.html.

13   **Forsaking fair**: *Com*, Sept. 25, 1901. **Railroad Day, Mabel Barnes**: *Express*, Sept. 29, 1901; *Com*, Sept. 25, 27, 28, 1901; Barnes, "Peeps," Vol. III, pp. 181–90.

14   **Lion-cage wedding**: *Express*, Sept. 29, 1901; *Courier*, Sept. 29, 1901.

15   **Illinois Day; hunger**: *Express*, Oct. 7, 1901. **Train accident**: *Express*, Oct. 7, 8, 1901; *Com*, Oct. 7, 1901. **Illinois speeches**: *Express*, Oct. 7, 8, 1901; *Com*, Oct. 7, 1901.

CHAPTER 8: FREEFALL

1   **Mood swings; bank panic**: *Courier*, Oct. 11, 1901; *Times*, Oct. 1, 1901; *Com*, Oct. 14, 1901.

2   **Czolgosz in prison**: *Brooklyn Daily Eagle*, Oct. 1, 1901; *Auburn Weekly Bulletin*, Oct. 8, 10, 11, 29, 1901; *Elmira Star Gazette*, Oct. 26, 1901; *Cortland Democrat*, Oct. 4, 1901; *News*, Oct. 15, 1901.

3   **Buffalo Day, brainstorming**: *Com*, Oct. 11, 19, 21, 1901; *Courier*, Oct. 20, 21, 1901.

4   **At the West Bay City Cooperage**: *Bay City Times-Press*, Oct. 3, 1901. **Leaving Bay City**: *Bay City Times-Press*, Oct. 8, 1901; *Daily Cataract Journal*, Oct. 17, 1901.

5   **Taylor's early life**: Anna Edson Taylor, *The Autobiography of Anna Edson Taylor* (printed booklet, n.d., n.p., Anna Taylor file, Niagara Falls Public Library), 2–3; *Express*, Oct. 21, 27, 1901. **Texas experiences**: Taylor, *Autobiography*, 3–4. **Later escapades**: Taylor, *Autobiography*, 4–8; *Express*, Oct. 21, 27, 1901; Whalen, *The Lady Who Conquered*, 1–17; Charles Carlin Parish, *Queen of the Mist: The Story of Annie Edson Taylor* (Interlaken, NY: Empire State Books, 1987), 31–44. On p. 40, Parish sums up the quandary of most of Taylor's biographers: "Where does

truth end and fantasy begin?" **Taylor in Asylum**: 1900 United States Federal Census, Traverse City, Michigan, accessed at interactive.ancestrylibrary.com/. **Evidence of residence in Texas**: Letters in post office for Mrs. David Taylor: *Galveston Daily News*, Aug. 3, 1879; lot sold to Anna E. Taylor for $5,000: *Fort Worth Daily Gazette*, March 24, 1887.

6   **Mrs. Odell's visit**: *Express*, Oct. 11, 1901. **Annie Taylor**: *Express*, Oct. 21, 1901. **Elite Women of Buffalo**: See Mary Rech Rockwell, "'Let Deeds Tell': Elite Women of Buffalo, 1880–1910," (PhD diss., State University of New York at Buffalo), esp. ch. 5. **Women Managers**: *Newark Sunday News*, Feb. 24, 1901; *Express*, Sept. 26, 1901; *Com*, Aug. 16, 1901; Director-General Buchanan to (Pan-American) Executive Committee, June 6, 1901, Buffalo History Museum Archives, Buchanan correspondence, Mss. C 64–6. **White City**: *Enq*, Feb. 1, 1901; *New York Evening Post*, Nov. 21, 1900. **Women as women**: *Harper's Weekly*, Aug. 3, 1901; *Com*, Oct. 8, 1901.

7   **The New Woman and mobility**: Virginia Scharff, *Women and the Coming of the Motor Age* (Albuquerque: University of New Mexico Press, 1999), 4; Amy G. Richter, *Home on the Rails: Women, the Railroad, and the Rise of Public Domesticity* (Chapel Hill: University of North Carolina Press, 2005), 45–55, 143–58; *Com*, Sept. 9, 1901. **Bicycles**: Ellen Gruber Garvey, "Reframing the Bicycle: Advertising-Supported Magazines and Scorching Women," *American Quarterly* 47 (March 1995): 67–69, 72. **Warnings**: Richter, *Home*, 45–55; Scharff, *Motor Age*, 25–26, 47, 71–72; Mona Domosh and Joni Seager, *Putting Women in Place: Feminist Geographers Make Sense of the World* (New York: Guilford Press, 2001), 124–25; Garvey, "Reframing," 70, 74–75, 80.

8   **Finding Truesdale**: Dwight Whalen, *The Lady Who Conquered Niagara* (Bailey Island, ME: EGA Books, 1991), 50–52. **Answering skeptics**: *Express*, Oct. 21, 1901; *Niagara Falls Journal*, Oct. 18, 1901; *Kalamazoo Gazette-News*, Oct. 16, 1901.

9   **Niagara River and Falls**: Ralph S. Tarr, "Physical Geography of New York State. Part VIII. The Great Lakes and Niagara," *Journal of the American Geographical Society of New York* 31 (1899): 324; D. W. Johnson, "Rate of Recession of Niagara Falls," *The American Naturalist* 41 (August 1907): 541–42, accessed at http://www.jstor.org/stable/2454830?seq=1#page_scan_tab_contents.

10   **Tourists, developers**: Linda Revie, *The Niagara Companion: Explorers, Artists, and Writers at the Falls, from Discovery through the Twentieth Century* (Waterloo, Ont.: Wilfrid Laurier University Press, 2003), 3–4; Patrick McGreevy, *Imagining Niagara: The Meaning and Making of Niagara Falls* (Amherst: University of Massachusetts Press, 1994), 36–37. **Performers**:

Pierre Berton, *Niagara: A History of the Falls* (Albany: State University of New York, 1992), 124–40, 189–90.

11 **Midleigh/Gardner**: *New York Times*, June 29, 1890. **Cat**: *Bay City Times-Press*, Oct. 19, 1901; *Express*, Oct. 21, 1901.

12 **Fairgrounds in disrepair**: *Express*, Oct. 25, 1901. **Shows closing**: *Com*, Oct. 24, 1901; *Express*, Oct. 24, 1901; *Courier*, Oct. 24, 1901. **Buchanan leaving**: *Com*, Oct. 25, 1901. **Pessimism and guarded optimism**: *Com*, Oct. 22, 26, 1901; *Express*, Oct. 21, 23, 1901. **Bostock oblivious**: *Express*, Oct. 23, 1901; *Com*, Oct. 7, 22, 23, 1901; *Courier*, Oct. 8, 1901.

13 **Suspicions, proportions, explanations**: *Express*, Oct. 21, 1901; *Niagara Falls Gazette*, Oct. 11, 1901. **Taylor's age**: Whalen, *The Lady Who Conquered Niagara*, 2.

14 **Impatience**: Whalen, *Lady Who Conquered*, 58; *Express*, Oct. 21, 1901. **Interviews**: Whalen, *Lady Who Conquered*, 59–61. **October 23 attempt**: *Courier*, Oct. 24, 1901; *Niagara Falls Gazette*, Oct. 23, 1901; *Express*, Oct. 25, 1901; [Niagara Falls] *Daily Cataract-Journal*, Oct. 24, 25, 1901; Whalen, *Lady Who Conquered*, 65–68.

15 **Annie Taylor goes over the falls**: *Daily Cataract-Journal*, Oct. 23, 24, 25, 1901; Dunlap, "Interview," Oct. 25, 1901; *Express*, Oct. 25, 1901; *Niagara Falls Review*, Oct. 26, 1901; Orrin E. Dunlap, Sr., "Plunging over Niagara Falls in a Barrel," typed recollection, 1920s, in Stunts and Stunters file, Niagara Falls Public Library; *News*, Oct. 25, 1901; *Fort Worth Register*, Oct. 28, 1901; Annie Edson Taylor, *Over the Falls: Annie Edson Taylor's Story of Her Life* (privately printed, 1902), 17; Whalen, *Lady Who Conquered*, 74–87.

CHAPTER 9: THE ESCAPE OF THE DOLL LADY

1 **Recovery; recollection; barrel**: *Daily Cataract-Journal*, Oct. 25, 1901. **Brain fever**: *Niagara Falls Gazette*, Oct. 26, 1901.

2 **Sales pitches**: *Com*, Oct. 28, 1901; *Courier*, Oct. 27, 1901; *Express*, Oct. 27, 28, 1901. **Midway mayor**: *Express*, Oct. 26, Nov. 1, 1901. **Chiquita**: *Com*, Oct. 26, 1901. **Shaking hands**: *Courier*, Aug. 4, 1901.

3 **Gifts**: *Com*, Oct. 28, 29, 1901; *Express*, Oct. 30, 1901; *News*, Oct. 28, 1901. **Hearsay**: *Courier*, Oct. 26, 1901; *Express*, Oct. 29, 1901. **Farewell Day**: *News*, Oct. 27, 1901; *Com*, Oct. 30, 1901.

4 **The poor and the Exposition**: *Express*, Oct. 27, 1901; *News*, Oct. 24, 29, 1901. **Farewell Day sham battle**: *Com*, Oct. 28, 29, 30, 1901; *Express*, Oct. 29, 1901; *Courier*, Oct. 27, 1901. **On Indian community**, see, for example, *Com*, July 13, Sept. 9, 1901; *Express*, July 17, 25, 1901. On Indian resistance at other fairs, see Josh Clough, "'Vanishing Indians?' Cultural Persistence on Display at the Omaha World's Fair of 1898,"

Great Plains Quarterly 25 (April 2005): 67–86, http://digitalcommons.unl
.edu/cgi/viewcontent.cgi?article=3472&context=greatplainsquarterly;
Nancy Egan, "Exhibiting Indigenous Peoples: Bolivians and the Chicago Fair of 1893," *Studies in Latin American Popular Culture* 28 (January
2010): 15–18. **Other events, including Taylor**: *Express,* Oct. 28, 29,
Nov. 1, 1901; *Courier,* Oct. 27, 1901.

5   **Assassination sites**: *Courier,* Oct. 28, 1901. **Czolgosz's death**:
Charles R. Skinner, "Story of McKinley's Assassination," *State
Service* 3 (Apr. 1919): 20–24, accessed at http://mckinleydeath
.com/documents/magazines/SService3-4.htm; *New York Times,*
Oct. 25, 1901; Carlos F. MacDonald, "The Execution of Czolgosz,"
Medical News 79 (Nov. 1901): 752–53, accessed at http://
mckinleydeath.com/documents/journals/MN79-19a.htm; *Auburn
Weekly Bulletin,* Nov. 1, 1901; "How Czolgosz Will Meet His
Death," *Black and White Budget* 6 (Oct. 1901): 138–39, accessed at
http://mckinleydeath.com/documents/magazines/BWB6-107a.htm;
Com, Oct. 29, 1901; *St. Louis Post-Dispatch,* Oct. 29, 1901.

6   **Electricity fears**: Jürgen Martschukat, "'The Art of Killing by Electric-
ity': The Sublime and the Electric Chair," *Journal of American History* 89
(December 2002): 911; *Express,* Oct. 28, 1901. **Current wars**: Jill Jonnes,
Empires of Light: Edison, Tesla, Westinghouse, and the Race to Electrify the World
(New York: Random House, 2004), 148–50, 201, 207; Gilbert King,
"Edison vs. Westinghouse: A Shocking Rivalry," Smithsonian.com, Oct.
11, 2011, accessed at http://www.smithsonianmag.com/history/edison-vs
-westinghouse-a-shocking-rivalry-102146036/?no-ist; Martschukat,"The
Art of Killing,"900, 915. **Kemmler's death**: Martschukat,"The Art of
Killing," 917–18; Jonnes, *Empires of Light,* 188–89.

7   **Packing up**: *Express,* Nov. 1, 1901. **Escape, marriage, capture**: *Erie
Daily Times,* Nov. 2, 1901; *Express,* Nov. 9, 1901; Bostock v. Woeckener, Testimony of Thomas Rochford; *Courier,* Nov. 2, 1901; *Kalamazoo Gazette,* Nov. 13, 1908. **Beating**: *Boston Daily Globe,* Jan. 4, 1902;
Bostock v. Woeckener, Testimony of Mrs. C. W. Page.

8   **Weather, good-byes**: *Express,* Oct. 29, Nov. 2, 3, 1901; *Com,* Nov. 2,
1901; *Courier,* Nov. 2, 1901.

9   **The Exposition as civilizer**: *Courier,* Oct. 14, 1901; *Express,* Oct. 3,
29, 1901. **Mixed feelings, resistance**: See Nancy Egan, "Exhibiting
Indigenous Peoples," 17. **Laughing Ben**: *Macon Telegraph,* Sept. 2, Nov.
11, 1901. **Filipino resistance**: *Courier,* Sept. 28, 1901. **Performer illness, deaths**: Rydell, *All the World's a Fair,* 150; *Express,* April 24, May
12, June 9, July 14, July 19, Nov. 10, 1901; *Com,* May 13, Sept. 9, 1901;
Courier, Aug. 9, 25, Oct. 22, 1901.

10   **Mabel visits for final time**: Barnes, "Peeps," Vol. III, pp. 195–99; *News,* Oct. 24, Oct. 27, 1901; *Express,* Nov. 2, 1901. **Carrie Nation**: *New York Times,* Sept. 9, 1901; *Express,* Nov. 2, 1901.

11   **Annie Taylor at the Exposition**: *Com,* Nov. 2, 1901; *Courier,* Nov. 3, 1901. **Compliments and criticism**: *Express,* Oct. 27, 1901; *Com,* Oct. 28, 1901; *Bay City Tribune,* Oct. 27, 1901; *Bay City Times-Press,* Nov. 5, 1901.

12   **Mabel's last night**: Barnes, "Peeps," Vol. III, pp. 197–99; *Express,* Nov. 3, 1901. **Scavengers, fireworks**: *Express,* Nov. 3, 1901.

13   **Temple of Music and lights out**: *Express,* Nov. 3, 1901; *Com,* Nov. 2, 1901. **Destruction**: *Express,* Nov. 3, 1901; *Courier,* Nov. 3, 1901.

14   **Bostock in charge**: *Kalamazoo Gazette,* Nov. 13, 1908. **Bostock spins the story**: *Erie* [PA] *Morning Dispatch,* Nov. 5, 6, 11, 1901; *Express,* Nov. 3, 1901.

CHAPTER 10: THE ELEPHANT

1   **Chiquita and courage**: *Kalamazoo Gazette,* Nov. 13, 1903. **Bostock in court**: *Express,* Nov. 9, 1901, Jan. 4, 1902; *Com,* Nov. 8, 1901; *News,* Nov. 8, 9, 1901; *Erie* [PA] *Express,* Nov. 9, 1901.

2   **Maitland's announcement**: *Enq,* Nov. 7, 1901. **Henry Mullen, Tina Caswell**: *Courier,* Nov. 1, 7, 1901; *Wilkes-Barre Times,* Oct. 31, 1901; *Enq,* Nov. 7, 1901; *Express,* Nov. 7, 1901.

3   **Jumbo's antics**: *Com,* July 29, 1901; *Courier,* July 24, 1901. **New, evil Jumbo**: *Newark* [Ohio] *Advocate,* Nov. 9, 1901; *Courier,* Nov. 7, 8, 1901; *Express,* Nov. 9, 1901; *Enq,* Nov. 8, 1901.

4   **Rajah and Trilby**: *New York Times,* April 13, 1901; *Com,* Oct. 12, 17, 1901; *Express,* Oct. 12, 17, 20, 1901. **Elephant deaths**: *Courier,* Nov. 7, 8, 1901; *Com,* Nov. 8, 1901; *The* [Fredericksburg, VA] *Free Lance,* June 12, 1900 . **Electrocution**: *Enq,* Nov. 7, 1901; *Courier,* Nov. 7, 8, 1901. **Jumbo advertisement**: *Courier,* Nov. 9, 1901; *Express,* Nov. 9, 1901; *Enq,* Nov. 8, 1901.

5   **Regrets**: *Courier,* Nov. 10, 1901; *Enq,* Nov. 7, 1901. **Old Bet**: G. G. Goodwin, "The First Living Elephant in America," *Journal of Mammalogy* 6 (November 1925): 257–61, plate 24. See also http://www.naturalhistorymag .com/editors_pick/1928_05-06_pick.html?page=2. **Original Jumbo**: *New York Times,* Sept. 17, 1885; Susan Wilson, "An Elephant's Tale: An Unadulterated and Relatively True Story Chronicling the Life, Death and Afterlife of Jumbo, Tufts' Illustrious Mascot," *Tufts online Magazine* 9 (Spring 2002), accessed at http://www.tufts.edu/alumni/magazine/ spring2002/jumbo.html. **Bostock's elephant shows**: *Com,* July 23, 1901; Aug. 3, 29, 1901; *Com,* Sept. 16, 1901.

6   **The Woeckener home**: *Erie Morning Dispatch,* Nov. 4, 6, 1901. **Tony's desperation**: *Erie Morning Dispatch,* Nov. 6, 1901; *Express,* Nov. 9, 1901; Jan. 5, 1902.

7   **Crowds arrive**: *Courier,* Nov. 10, 1901; *News,* Nov. 3, 1901. **Electricity details**: *Express,* Nov. 9, 1901; *Courier,* Nov. 9, 1901. **Protests**: *Express,* Nov. 9, 1901.

8   **Mary Lord, Henry Bergh, mules**: Margaret F. Rochester, *Lest We Forget: Historical Sketch of the Erie County Society for the Prevention of Cruelty to Animals* (Buffalo: privately printed, 1916); *Buffalo Evening News,* Sept. 15, 1965. **Mary Lord, Millard Fillmore**: George J. Bryan, *Biographies of Attorney-General George P. Barker, John C. Lord, D.D., Mrs. John C. Lord, and William G. Bryan, Esq.* (Buffalo: The Courier Company, 1886), 150–54; *Buffalo Evening News Magazine,* June 8, 1968; Frank Severance, ed., *Publications of the Buffalo Historical Society* XI (New York: Buffalo Historical Society, 1907); *Buffalo Historical Society: Annual Report of the Board of Managers for the Year 1898* (Buffalo: Baker, Jones & Co., printers, 1899), 75.

9   **Humane Society at the Exposition**: *Annual Report: Erie County Society for the Prevention of Cruelty to Animals* (Buffalo: Turner & Porter Printers, 1901), 8–10, 20, 30–33; *News,* May 29, 1901; *Com,* May 6, 1901; *Courier,* Nov. 6, 1901.

10 **Social control and animal rescue**: See Kathleen Kete, ed., *A Cultural History of Animals in the Age of Empire* (Oxford: Berg, 2007), 2–4. **Bostock cruelty charges, England**: *Manchester Weekly Times,* Aug. 29, 1890; *North-Eastern Daily Gazette,* March 17, Nov. 4, 1891; *York Herald,* March 4, 1891; *Coventry Evening Telegraph,* April 7, 1892.

11 **The attempted killing**: *Express,* Nov. 10, 1901; *Courier,* Nov. 10, 1901; *Charlotte Daily Observer,* Nov. 12, 1901. **Bostock's new plans**: *Courier,* Nov. 10, 1901. **Jumbo II in Boston**: *Boston Daily Globe,* Nov. 29, 1901.

12 **Where is Chiquita?**: *Erie Morning Dispatch,* Nov. 18, 1901. **Boston court battle**: *Erie Morning Dispatch,* Dec. 14, 1901; Jan. 4, 8, 10, 1902; September 16, 1902; Bostock v. Espiridiona Alice Cenda Woeckener, Equity Case No. 27, November Term, 1902, Circuit Court of the United States for the Western District of Pennsylvania, Philadelphia Archives.

13 **Contract**: *Erie Morning Dispatch,* Sept. 16, 1902; Bostock v. Woeckener. **Bostock's offer**: *Times* Jan. 16, 1902; *Erie Morning Dispatch,* Jan. 11, 1902. **Chiquita goes to Glasgow**: *Times,* Jan. 16, 1902; *Glasgow Daily Record & Daily Mail,* Jan. 23, 1902. **Tony relents**: *Erie Morning Dispatch,* Feb. 24, March 3, 1902.

14 **Bostock insurance**: *Erie Morning Dispatch,* March 31, 1902; *Boston Daily Globe,* March 27, 1902. **Under the rule of the Badgers**: Bostock v.

Woeckener; *Erie Morning Dispatch,* Aug. 28, 1902; *Rome Daily Sentinel,* Aug. 29, 1902.

15   **Escape from Elgin**: *Erie Morning Dispatch,* Aug. 28, 1902. **Chiquita in Erie**: *Erie Morning Dispatch,* Aug. 28, 29, 1902. **Bostock plots, sues**: [Batavia, NY] *Daily News,* Sept. 14, 1904; Bostock v Woeckener, Exhibit "C"; Al Stencell, "Frank Bostock in America," in National Fairground Archive, Sheffield University, accessed August 13, 2014, at http://www.nfa.dept.shef.ac.uk/jungle/index1a1.html; *Erie Morning Dispatch,* Sept. 16, 1902; *Elgin* [IL] *Daily Courier,* Aug. 27, 28, 1902. **Testimonies**: Bostock v. Woeckener. **Case closed, Chiquita advertised**: *Erie Morning Dispatch,* Jan. 5, 17, 1903; Bostock v. Woeckener.

CHAPTER 11: THE TIMEKEEPERS

1   **Jumbo II tours, arrives in Cleveland**: *Cleveland Plain Dealer,* June 24, 25, 1902. **Jumbo at Manhattan Beach**: *Cleveland Plain Dealer,* June 24, 25, 26, 29, July 3, Aug. 30, 1902. **Left with Cleveland sheriff**: *Cleveland Plain Dealer,* Oct. 4, 14, Nov. 16, 1902. **Bostock takes charge again**: *Cleveland Plain Dealer,* Nov. 16, 1902; *Charlotte Daily Observer,* Nov. 17, 1902. **Jumbo II dies**: *Cleveland Plain Dealer,* Nov. 16, 18, 1902.

2   **Bostock, Blondin, Bonavita**: *Cambridge* [MA] *Tribune,* Sept. 12, 1908; *New-York Tribune,* March 21, 1917; *New York Times,* Oct. 9, 1912. **Bostock's death, funeral, tributes**: *New York Times,* Oct. 9, 1912; Al Stencell, "Frank Bostock in America," *The World's Fair,* Oct. 12, 19, 1912, accessed March 22, 2015, at http://www.nfa.dept.shef.ac.uk/jungle/index1a4.html.

3   **Topsy**: *New York Herald,* Jan. 5, 1903; *New York Press,* Jan. 5, 1903. **Roosevelt**: Theodore Roosevelt, *African Game Trails: An Account of the African Wanderings of an American Hunter-Naturalist* (New York: Charles Scribner's Sons, 1910), 2, 9, 29, 283–84; 290–92; on hunting and imperialism, see also Joseph Sramek, "'Face Him Like a Briton': Tiger Hunting, Imperialism, and British Masculinity in Colonial India, 1800–1875," *Victorian Studies* 48 (June 2006): 659–80. **Retiring the herd**: See http://news.nationalgeographic.com/news/2015/03/150305-ringling-bros-retires-asian-elephants-barnum-bailey/.

4   **Chiquita and Tony's plans, pregnancy**: *Erie Morning Dispatch,* Jan. 5, 1903; [Batavia, NY] *Daily News,* Oct. 12, 1903; [Syracuse NY] *Evening Telegram,* Oct. 18, 1903. **Later tours**: *Express,* Aug. 29, 1907; *Courier,* July 3, 1910; *Billboard,* July 28, 1906. **Chiquita's death**: *News,* April 17, 1928.

5   **On "midget" circuses**: Robert W. Rydell, John E. Findling, Kimberly D. Pelle, *Fair America: World's Fairs in the United States* (Washington: Smithsonian Books, 2000), 82–90. **Little People, rights and respect**:

Robert Bogdan, *Freak Show: Presenting Human Oddities for Amusement and Profit* (Chicago: University of Chicago Press, 1988), 30–31, 64–66; Rosemarie Garland Thomson, *Extraordinary Bodies: Figuring Physical Disability in American Culture and Literature* (New York: Columbia University Press, 1997), 22, 74–75, 78–80.

6 **Attendance figures**: Isabel Vaughan James lists the final Pan-American admission figure as 8,120,048, with 5,306,859 as paid admissions. See Vaughan James, *The Pan-American Exposition*, 13; *Courier*, Nov. 2, 1901. **Charleston**: Anthony Chibbaro, *The Charleston Exposition* (Charleston, SC: Arcadia Publishing, 2001), 7–8, 56, 65, 76–77.

7 **Taylor exhaustion, age challenges**: *Bay City Tribune*, Oct. 30, Nov. 8, 9, 10, 1901; Whalen, *The Lady Who Conquered*, 117–22, 133; *Niagara Falls Gazette*, March 31, 1902.

8 **Sanatorium, tours, struggles**: *Bay City Times-Press*, Nov. 15, 1901; Whalen, *The Lady Who Conquered*, 120–25, 128–30, 132; *Bay City Tribune*, Feb. 14, 1902. **Stolen barrel, despair**: Whalen, *The Lady Who Conquered*, 134–35, 140–42; Annie Taylor to Frank Tanner, March 26, 1902, in the *Niagara Falls Gazette*, March 31, 1902. The play, with "the original barrel," was advertised periodically from August 1902 to at least 1906. See the *New York Clipper*, Aug. 30, 1902 and *The* [Rock Island, IL] *Argus*, March 5, 1906.

9 **Barrel recovery, new manager**: *Bay City Times-Press*, Aug. 20, 1902; Whalen, *The Lady Who Conquered*, 150–53; *Trenton Evening Times*, Oct. 9, 1902. **Barrel stolen again**: *Trenton Evening Times*, Oct. 9, 1902; *Cleveland Plain Dealer*, Oct. 10, 1902; Sept. 17, 1903; *Grand Forks Daily Herald*, Aug. 30, 1903; Whalen, 150–53. **Taylor in later life**: *Niagara Falls Gazette*, July 31, 1903; *Niagara Falls Journal*, July 26, 1911; Nov. 30, 1914; Whalen, 158–59, 160–64. **Almshouse**: *Niagara Falls Gazette*, March 4, 7, 1921; *Lockport Union-Sun and Journal*, April 30, 1921; Whalen, 166–69; 172–73, 176.

10 **Resenting Taylor's feat**: *Charlotte Daily Observer*, Sept. 20, 1902.

11 **Charleston exposition**: Chibbaro, *Charleston Exposition*, 56; Sara S. Cromwell, "Fair Treatment? African-American Presence at International Expositions in the South, 1884–1902" (MA thesis, Wake Forest University, 2010), 127–35. **Ben Ellington**: *Macon Telegraph*, Nov. 11, Sept. 12, 1901; *Atlanta Constitution*, April 30, 1905.

12 **Parker lectures**: *Washington Post*, Dec. 22, 1901; *The Colored American*, Nov. 30, 1901; *Cleveland Gazette*, March 23, 1907. **Parker tours; Hanna; Ida McKinley**: *Washington Post*, Dec. 22, 1901; *Topeka Plaindealer*, Jan. 3, 1902; *The Freeman*, Jan. 4, 1902; April 25, 1903; *The Worcester Spy*, Sept. 17, 1902; *Brooklyn Eagle*, Aug. 20, 1902; *The Colored*

American, Nov. 30, 1901; March 28, 1903; *New-York Tribune,* March 23, 1907; *New York Times,* March 24, 1907; James B. Parker to Ida McKinley, in Anthony, *Ida McKinley,* ebook location 6053–6057; 6080. **Parker hospitalized, dies:** *Washington Post,* March 27, 1908; *Broad Axe,* April 11, 1908; *Cleveland Gazette,* April 11, 1908.

13 **Mary Talbert and the Niagara Movement:** Peggy Brooks-Bertram and Barbara Seals Nevergold, *Uncrowned Queens: African American Women Community Builders* (Buffalo: Uncrowned Queens Publishing, 2005), 164–81. **Du Bois:** Brooks-Bertram and Nevergold, *Uncrowned Queens,* vii.

14 **The Goddess of Light falls:** *News,* July 2, 1902; *Courier,* July 13, 1902; *Com,* July 1, 2, 1902; *Express,* July 2, 1902. **The fair goes down:** *Courier,* July 13, 1902; *Express,* June 14, 1903; *Express,* Jan. 5, 1902. **Can't it be saved?** *Express,* Oct. 21, 27, Nov. 17, 1901; *Courier,* Oct. 2, 1901; *News,* Oct. 21, 1901.

15 **The balance sheet:** Buchanan, *Pan American Exposition Report,* p. 8; *Com,* June 24, 1902. **Blaming the assassination:** *Courier,* Oct. 29, 1901; *News,* Dec. 1, 1901. **Blaming other circumstances:** *Express,* July 5, 1902; *Com,* July 8, 1901; *Express,* Nov. 2, 1901; Buchanan, *Pan American Exposition Report,* p. 6; *Com,* Nov. 2, 1901.

16 **John Milburn is proud:** *Courier,* Oct. 26, 1901. **Pan Americanism accomplished:** *Courier,* Nov. 2, 1901. **The eyes of the world:** *Com,* Nov. 2, 1901; *Courier,* Oct. 26, Nov. 2, 1901. **Washington, New York help out:** *New York Times,* July 1, 1902.

17 **Structures' second life:** *Express,* Aug. 8, 1902; *Courier,* Aug. 11, 1901. While the *Courier* announced in August 1902 that the "McKinley" flooring was going to Washington to the National History Museum, the *Buffalo News* announced on September 14, 1902, that plans had changed. **New York State Building:** http://www.preservationbuffaloniagara.org/buildings-and-sites/buildings-catalog/location:buffalo-historical-society/. **Mabel Barnes:** *Buffalo Courier-Gazette,* Jan. 24, 1946, and http://www.buffalolib.org/sites/default/files/pdf/genealogy/subject-guides/Births%20Deaths%20and%20Marriages%20Found%20in%20Local%20Publications.pdf.

18 **Buchanan:** "William I. Buchanan, Pan-American Diplomat," *Bulletin of the International Union of the American Republics* 29 (October–December 1909): 835–37; Harold F. Peterson, *Diplomat of the Americas: A Biography of William I. Buchanan, 1852–1909* (Albany: SUNY Press, 1977). **Milburn house:** *Annual Report of the Board of Police of the City of Buffalo* (Buffalo: Wenborne-Sumner Co., 1902), 18; *Courier,* Oct. 21, 1901. **Milburn subsequent life and death:** *New York Times,* Aug. 12, 1930;

Susan Eck, "The Milburns and their Famous Home: 1168 Delaware Avenue," http://wnyheritagepress.org/photos_week_2009/mckinley _marker/milburn_house/milburn_house.htm. **Diehl:** *News,* March 13, 1902; *Express,* Feb. 21, 1918; *Times,* Feb. 15, 1919.

19 **Roosevelt on conservation:** Jeffrey Salmon, "'With Utter Disregard of Pain and Woe,': Theodore Roosevelt on Conservation and Nature," in Charles T. Rubin, ed., *Conservation Reconsidered: Nature, Virtue, and American Liberal Democracy* (Lanham, MD: Rowman & Littlefield, 2000), 49. **Roosevelt shot:** Edmund Morris, *The Rise of Theodore Roosevelt* (New York: Modern Library, 2001), p. xxxi; Patricia O'Toole, "Assassination Foiled," Smithsonian 43 (Nov. 2012), accessed on May 26, 2015 at http://web.b.ebscohost.com.lprx.bates.edu/ehost/detail/detail?sid =b2086474-c6f7-4c71-8f7e-5b650711eaa1%40sessionmgr114&vid =7&hid=124&bdata=JnNpdGU9ZWhvc3QtbGl2ZQ%3d%3d#d b=a9h&AN=83097307), and Rauchway, *Murdering McKinley,* 197. **Roosevelt's speech:** http://www.businessinsider.com/heres-the -famous-populist-speech-teddy-roosevelt-gave-right-after-getting-shot -2011-10#ixzz3dWyP6oRl. **Goldman:** "Radical Comment on the President's Assassination," *Literary Digest* 21 (Sept. 1901): 336–37.

20 **Ida McKinley in mourning:** Anthony, *Ida McKinley,* ebook, ch. 17. **Improving health, death:** Anthony, *Ida McKinley,* ebook, ch. 18.

21 **Buffalo steel; Exposition results; Buffalo booming:** *Courier,* July 13, 1902; *World's Fair Bulletin* 4 (December 1902): 16; Mark Goldman, *City on the Edge: Buffalo, New York* (Amherst, NY: Prometheus Books, 2007), 133–35. **Needing hustle:** Sarah Elvins, *Sales & Celebrations: Retailing and Regional Identity in Western New York State, 1920–1940* (Athens: Ohio University Press, 2004), 39–40. **Buffalo at midcentury:** Elvins, *Sales & Celebrations,* 48; Goldman, *City on the Edge,* 99–129, 162–68; *Courier Express,* June 5, 1963.

22 **Saint Lawrence Seaway:** Goldman, *City on the Edge,* 152–53, 315. **Layoffs, plant closings:** Goldman, *City on the Edge,* 308–11, 314; Goldman, *City on the Lake,* 177; Carol J. Loomis, "The Sinking of Bethlehem Steel," *Fortune Magazine,* April 5, 2004; "82 Years of History in Lackawanna," *New York Times,* December 28, 1982; Ardith Hilliard, David Venditta, eds., *Forging America: The History of Bethlehem Steel* (Allentown, PA: *The Morning Call,* 2010), accessed September 13, 2015, at http://www.mcall.com/all-bethsteel-printingchapter-8-htmlstory.html.

23 **Population:** Census: https://www.census.gov/statab/hist/HS-07.pdf. **Centennial meanings:** See Michael Frisch's thoughtful essay, "Prismatics, Multivalence, and Other Riffs on the Millennial Moment: Presidential Address to the American Studies Association," *American Quarterly*

53 (2001): 193–231. **On the virtual Pan-American**, see Susan Eck's pioneering website, "Doing the Pan," at http://panam1901.org/. This site offers a visit to the exposition through a plethora of firsthand accounts and other primary sources; see also the University at Buffalo's website on the Exposition, with essays on topics from architecture to African American history: http://library.buffalo.edu/pan-am/exposition/. **Bestselling novel**: Lauren Belfer, *City of Light* (New York: Dial Press, 1999). **Other important centennial work**: Kerry L. Grant, *The Rainbow City: Celebrating Light, Color, and Architecture at the Pan-American Exposition, Buffalo 1901* (Buffalo: Canisius College Press, 2001); Leary and Sholes, *Buffalo's Pan-American*. **The Buffalo commemoration**: Patrick Klinck, "As a Century Turns: Finding the Light Then and Now—The Pan-American Exposition," *UB Today* (Winter 2000): 16–17. **Gazing back**: Randal C. Archibold, "Buffalo Gazes Back to a Time When Fortune Shone: Much-Maligned City Celebrates the Glory of a Century Ago," *New York Times*, Sept. 7, 2001. **New York and Buffalo**: *News,* Aug. 30, 1901, quoting the *Chicago Record-Herald;* Mitchell Moss, Op-Ed, *New York Times,* March 28, 2010; see also the cartoon by Tom Toro in *The New Yorker,* Dec. 5, 2011. **Buffalo rising**: Author conversations with Leslie Zemsky, Mark Goldman, Steve Bell, Michael Frisch. See also Brian Hayden, "Buffalo's Grain Elevators Reimagined," accessed November 4, 2015, at http://www.visitbuffaloniagara.com/; and *New York Times,* Oct. 25, 2015; also, http://www.elkharttruth.com/news/national/2015/10/04/Buffalo-s-industrial-cathedrals-slowly-finding-new-life.html. **Legacies, vestiges of Rainbow City**: Mark Lozo to author, email correspondence, Sept. 14, 2015; Susan Eck to author, email correspondence, Sept. 30, 2015. It should also be noted that the Buffalo History Museum, along with Forest Lawn Cemetery and Explore Buffalo, offer popular Pan-American–focused tours, and that the museum's Forest Avenue Resource Center opens its extensive Pan-American exhibit on regular occasions.

24 **World's fairs**: Munro, "Investigating World's Fairs," 80. **Expositions in flux**: Rydell, Findling, Pelle, *Fair America,* 118; *Ocala* [WA] *StarBanner,* May 16, 1974.

Selected Bibliography

SECONDARY SOURCES

Adelson, Betty M. "Dwarfs: The Changing Lives of Archetypal 'Curiosities'—and Echoes." *Disability Studies Quarterly* 25, no. 3 (Summer 2005), http://dsq-sds.org/article/view/576/753. Accessed July 23, 2014.

Aldrich, William S. "Mechanical and Electrical Features of the Pan-American Exposition." *The Engineering Magazine* 21 (April to September 1901): 839–42.

Anthony, Carl Sferrazza. *Ida McKinley: The Turn of the Century First Lady Through War, Assassination, and Secret Disability.* Kent, OH: Kent State University Press, 2012.

Barry, Richard H. *Snap Shots on the Midway of the Pan-Am Expo.* Buffalo: Robert Allan Reid, 1901.

Beers, Diane L. *For the Prevention of Cruelty: The History and Legacy of Animal Rights Activism.* Athens, OH: Ohio University Press, 2006.

Bogdan, Robert. *Freak Show: Presenting Human Oddities for Amusement and Profit.* Chicago: University of Chicago Press, 1988.

Briggs, L. Vernon. *The Manner of Man That Kills: Spencer—Czolgosz—Richeson.* Boston: Richard G. Badger, 1921.

Brooks-Bertram, Peggy, and Barbara Seals Nevergold. *Uncrowned Queens: African American Women Community Builders.* Buffalo: Uncrowned Queens Publishing, 2005.

"The Case of the Late President McKinley." *British Medical Journal* 2 (November 1901): 1348–50.

Chibbaro, Anthony. *The Charleston Exposition.* Charleston, SC: Arcadia Publishing, 2001.

Cromwell, Sara S. "Fair Treatment? African-American Presence at International Expositions in the South, 1884–1902." MA thesis, Wake Forest University, 2010.

DeToledo, John C., Bruno B. DeToledo, and Meredith Lowe. "The Epilepsy of First Lady Ida Saxton McKinley." *Southern Medical Journal* 90, no. 3 (March 2000): 267–71.

Du Bois, W. E. B. "The Negro Exhibit." *American Monthly Review of Reviews* 22 (November 1990).

Dussel, Ines. "Between Exoticism and Universalism: Educational Sections in Latin American Participation at International Exhibitions, 1860–1900." *Paedagogica Historica* 47(October 2011): 601–17.

Egan, Nancy. "Exhibiting Indigenous Peoples: Bolivians and the Chicago Fair of 1893." *Studies in Latin American Popular Culture* 28 (January 1, 2010): 6–24.

Elvins, Sarah. *Sales & Celebrations: Retailing and Regional Identity in Western New York State, 1920–1940*. Athens, OH: Ohio University Press, 2004.

Everett, Marshall. *Complete Life of William McKinley and Story of his Assassination*. Chicago: C. W. Stanton, 1901. http://babel.hathitrust.org/cgi/pt?id=uc2.ark:/13960/t5fb4x97r;view=1up;seq=7.

Ferguson, Kathy. "Discourses of Danger: Locating Emma Goldman." *Political Theory* 36 (October 2008): 735–61.

Fernandez-Bravo, Alvaro. "Ambivalent Argentina: Nationalism, Exoticism, and Latin Americanism at the 1889 Paris Universal Exposition." *Nepantla: Views from South* 2 (January 2001): 115–39.

Fine, Sidney. "Anarchism and the Assassination of McKinley." *American Historical Review* 60 (July 1955): 777–79.

Fisher, Jack C. *Stolen Glory: The McKinley Assassination*. La Jolla, CA: Alamar Books, 2001.

Frisch, Michael. "Prismatics, Multivalence, and Other Riffs on the Millennial Moment: Presidential Address to the American Studies Association, 13 October 2000." *American Quarterly* 53 (2001): 193–231.

Ghartey-Tagoe, Amma Y. "The Battle Before the Souls of Black Folk: Black Performance in the 1901 Pan-American Exposition." PhD diss., New York University, 2009.

Gilbert, James. *Whose Fair? Experience, Memory, and the History of the Great St. Louis Exposition*. Chicago: University of Chicago Press, 2009.

Goldman, Mark. *City on the Edge: Buffalo, New York*. Amherst, NY: Prometheus Books, 2007.

Goldman, Mark. *City on the Lake: The Challenge of Change in Buffalo, New York*. Amherst, NY: Prometheus Books, 1990.

Goldman, Mark. *High Hopes: The Rise and Decline of Buffalo, New York*. Albany: State University of New York Press, 1983.

Grant, Kerry L. *The Rainbow City: Celebrating Light, Color, and Architecture at the Pan American Exposition, Buffalo 1901*. Buffalo: Canisius College Press, 2001.

James, Isabel Vaughan. *The Pan-American Exposition (Adventures in Western New York History)*. Buffalo: Buffalo and Erie County Historical Society, 1961.

Jayes, Janice Lee. *The Illusion of Ignorance: Constructing the American Encounter with Mexico, 1877–1920*. Lanham, MD: University Press of America, 2011.

Jonnes, Jill. *Empires of Light: Edison, Tesla, Westinghouse, and the Race to Electrify the World*. New York: Random House, 2004.

Kachun, Mitch. "'Big Jim' Parker and the Assassination of William McKinley: Patriotism, Nativism, Anarchism, and the Struggle for African American Citizenship." *The Journal of the Gilded Age and Progressive Era* 9 (January 2010): 93–116.

Kete, Kathleen. *A Cultural History of Animals in the Age of Empire*. London: Bloomsbury, 2011.

Kootin, Amma Y. Ghartey-Tagoe. "Lessons in Blackbody Minstrelsy." *The Drama Review* 57, no. 2 (Summer 2013): 102–22.

Larson, Erik. *The Devil in the White City: Murder, Magic, and Madness at the Fair That Changed America*. New York: Crown, 2003.

Leary, Thomas and Elizabeth Sholes. *Buffalo's Pan-American Exposition*. Charleston, SC: Arcadia Publishing, 1998.

Loos, William H., Ami M. Savigny, Robert M. Gurn, and Lillian S. Williams. *The Forgotten "Negro Exhibit": African American Involvement in Buffalo's Pan-American Exposition, 1901*. Buffalo: Buffalo and Erie County Public Library and Library Foundation of Buffalo and Erie County, 2001.

MacDonald, Carlos F. "The Trial, Execution, Autopsy and Mental Status of Leon F. Czolgosz, Alias Fred Nieman, the Assassin of President McKinley." *American Journal of Insanity* 58 (January 1902): 369–86. http://mckinleydeath.com/documents/journals/AJI58-3b.htm. Accessed December 14, 2015.

Martschukat, Jürgen. "'The Art of Killing by Electricity': The Sublime and the Electric Chair." *The Journal of American History* 89, no. 3 (December 2002): 900–21.

McHenry, Katherine V. "Color Scheme at the Pan American." *Brush and Pencil* 8 (June 1901): 151–56.

Moore, Sarah J. "Mapping Empire in Omaha and Buffalo: World's Fairs and The Spanish-American War." *Bilingual Review/Editorial Bilingüe* 25, no. 1 (January–April 2000): 111–26.

Morgan, Howard Wayne. *William McKinley and His America*. Kent, OH: Kent State University Press: 2003.

Munro, Lisa. "Investigating World's Fairs: An Historiography." *Studies in Latin American Popular Culture* 28 (2010): 80–94.

Parker, LeRoy. "The Trial of the Anarchist Murderer Czolgosz." *Yale Law Journal* 11, no. 2 (December 1901): 80–94.

Peavler, David J. "African Americans in Omaha and the 1898 Trans-Mississippi and International Exposition." *Journal of African American History* 93 (Summer 2008): 337–61.

Perez, Louis A., Jr. *Cuba in the American Imagination: Metaphor and the Imperial Ethos.* Chapel Hill: University of North Carolina Press, 2008.

Rauchway, Eric. *Murdering McKinley: The Making of Teddy Roosevelt's America.* New York: Hill and Wang, 2003.

Rinehart, Melissa. "To Hell with the Wigs!: Native American Representation and Resistance at the World's Columbian Exposition." *American Indian Quarterly* 36, no. 4 (Fall 2012): 403–42.

Rixey, Presley M., Matthew D. Mann, Herman Mynter, Roswell Park, Eugene Wasdin, Charles McBurney, and Charles G. Stockton. "The Case of President McKinley." *Medical Record* 60 (October 1901): 601–6. http://mckinleydeath.com/documents/journals/MR60-16dp.htm. Accessed October 20, 2015.

Rockwell, Mary Rech. "'Let Deeds Tell': Elite Women of Buffalo, 1880–1910." PhD diss., State University of New York at Buffalo.

Rydell, Robert W. "'Darkest Africa': African Shows at America's World's Fairs, 1893–1940." In *Africans on Stage*, ed. Bernth Lindfors. Bloomington: Indiana University Press, 2010.

Rydell, Robert W., John E. Findling, and Kimberly D. Pelle. *Fair America: World's Fairs in the United States.* Washington: Smithsonian Books, 2000.

Rydell, Robert W. *All the World's a Fair: Visions of Empire at American International Expositions, 1876–1916.* Chicago: University of Chicago Press, 1984.

Rydell, Robert W., and Nancy E. Gwinn, eds. *Fair Representations: World's Fairs and the Modern World.* Amsterdam: Vanderbilt University Press, 1994.

Ryder, Richard D. *Animal Revolutions: Changing Attitudes Toward Speciesism.* Oxford: Berg Publishers, 2000.

Severance, Frank H., ed. *Publications of the Buffalo Historical Society, Volume XI.* New York: Buffalo Historical Society, 1907.

Severance, Frank H., ed. *Millard Fillmore Papers, Volume II.* New York: Buffalo Historical Society, 1907.

Scharff, Virginia. *Taking the Wheel: Women and the Coming of the Motor Age.* New York: Free Press, 1991.

Sherman, Matthew C. "Protecting the First Citizen of the Republic: Presidential Security from Thomas Jefferson to Theodore Roosevelt." PhD diss., Saint Louis University, 2011.

Shupe, Kevin D. "Geronimo Escapes: Envisioning Indianness in Modern America." PhD diss., George Mason University, 2011.

Stencell, Al. "Frank Bostock in America." National Fairground Archive, Sheffield University, UK. http://www.nfa.dept.shef.ac.uk/jungle/index1a1.html. Accessed August 13, 2014.

"The Surgical and Medical Treatment of President McKinley." *Journal of Medicine and Science* 7 (October 1901): 389–90. http://mckinleydeath.com/documents/journals/JMSc7-11bp.htm. Accessed December 15, 2015.

Tenorio-Trillo, Mauricio. *Mexico at the World's Fairs: Crafting a Modern Nation.* Berkeley: University of California Press, 1996.

Thomson, Rosemarie Garland. *Extraordinary Bodies: Figuring Physical Disability in American Culture and Literature.* New York: Columbia University Press, 1997.

Thomson, Rosemarie Garland, ed. *Freakery: Cultural Spectacles of the Extraordinary Body.* New York: New York University Press, 1996.

Turner, C. Y. "The Pan-American Color Scheme." *The Independent* 53 (April 1901): 948–49.

Vials, Chris. "The Despotism of the Popular: Anarchy and Leon Czolgosz at the Turn of the Century." *Americana: The Journal of American Popular Culture* 3 (Fall 2004). http://www.americanpopularculture.com/journals/articles/fall_2004/vials.htm. Accessed October 8, 2015.

Walden, Keith. *Becoming Modern in Toronto: The Industrial Exhibition and the Shaping of Late Victorian Culture.* Toronto: Toronto University Press, 1997.

Whalen, Dwight. *The Lady Who Conquered Niagara: The Annie Edson Taylor Story.* Bailey Island, ME: EGA Books, 1990.

PRIMARY SOURCES

Annual Report of the Board of Police of the City of Buffalo. Buffalo: Wenborne-Sumner Co., 1902.

Annual Report of the Erie County Society for the Prevention of Cruelty to Animals. Buffalo: Turner & Porter, 1901.

Baird, Frank. Diary. Private Collection.

Barnes, Mabel E. "Peeps at the Pan-American. An account of Personal Visits in the Summer of 1901 from Notes jotted down on the Spot and put in Permanent form during Fourteen Years." Buffalo: Buffalo History Museum Archives, Mss. W-119.

Bennitt, Mark, *The Pan-American Exposition and How to See It.* Buffalo: The Goff Company, 1901.

Buchanan, William I. *Pan American Exposition: Report of William I. Buchanan, Director-General,* 1902. Buffalo: Special Collections, Buffalo and Erie County Public Library.

Channing, Walter, M.D., Papers. Massachusetts Historical Society, Boston.

Dunlap, Orrin E. "Interview with Mrs. Taylor." October 25, 1901. Unpublished manuscript. Stunts and Stunters file. Local History Collection, Niagara Falls (NY) Public Library.

Frank C. Bostock's Grand Zoological Congress and Trained Animal Arena. Buffalo: Courier Co., 1901.

Pan American Exposition Buffalo: Its Purpose and Plan. Buffalo: Pan-American Exposition Company, 1901. http://digital.hagley.org/cdm/ref/collection/p268001coll12/id/6749. Accessed October 1, 2015.

Park, Roswell. "Reminiscences of McKinley Week." Buffalo: Buffalo History Museum Archives, Mss. A00-390.

Taylor, Annie Edson. Over the Falls: Annie Edson Taylor's Story of Her Life. Privately printed, 1902.

Wells, Ida B., ed. "The Reason Why the Colored American is not in the World's Columbian Exposition." Chicago, 1893. Accessed December 14, 2015. http://www.loc.gov/resource/mfd.25023/?sp=1.

Williams, Edward T. "Martha E. Wagenfuhrer, 'Maid of the Rapids.'" Unpublished manuscript. Stunts and Stunters file. Local History Collection, Niagara Falls [NY] Public Library.

Williams, Edward T. "Maud Willard Meets Death in Whirlpool." Unpublished manuscript. Stunts and Stunters file. Local History Collection, Niagara Falls [NY] Public Library.

NEWSPAPERS

Bay City [Michigan] Times-Press, 1901–1902
Bay City [Michigan] Tribune, 1901
Buffalo Commercial, 1899–1902
Buffalo Courier, 1899–1902
Buffalo Enquirer, 1899–1902
Buffalo Evening News, 1899–1903
Buffalo Morning Express, 1899–1903
Buffalo Times, 1900–1901
Cleveland Plain Dealer, 1902
Erie [Pennsylvania] Morning Dispatch, 1901–1903
Niagara Falls Gazette, 1901–1902; 1921
Niagara Falls Journal, 1901, 1902, 1914
Pan-American Herald, 1899–1901
Pan American Magazine, 1900–1901

TRIAL TRANSCRIPTS

Frank C. Bostock v. Espiridiona Alice Cenda Woeckener, Equity Case No. 27, November Term, 1902, Circuit Court of the United States for the Western District of Pennsylvania. Philadelphia Archives.

The People of the State of New York v. Leon F. Czolgosz. Unpublished trial transcript, September 23–24, 26, 1901. Accessed December 12, 2015. http://mckinleydeath.com/documents/govdocs/transcript.htm.

WEBSITES

Eck, Susan. "Doing the Pan." http://panam1901.org/. Accessed December 12, 2015.

"Mckinley Assassination Ink: A Documentary History of William McKinley's Assassination." http://mckinleydeath.com/. Accessed December 12, 2015.

Old Fulton New York Post Cards. http://fultonhistory.com/. Accessed December 12, 2015.

"Pan-American Exposition of 1901." University at Buffalo. library.buffalo.edu/pan-am/. Accessed January 1, 2016.

"Pan-American Exposition Scrapbooks." Buffalo and Erie County Public Library/New York Heritage Digital Collections. http://www.nyheritage.org/collections/pan-american-exposition-scrapbooks. Accessed January 1, 2016.

Index

Note: Page numbers after 282 refer to notes.